Shelf-life Studies in Food Science

SHELF-LIFE STUDIES IN FOOD SCIENCE

Edited by:

Laghulkar Anita Marotirao

Delve
PUBLISHING

www.delvepublishing.com

Shelf-life Studies in Food Science

Laghulkar Anita Marotirao

Delve Publishing
224 Shoreacres Road
Burlington, ON L7L 2H2
Canada
www.delvepublishing.com
Email: orders@arclereducation.com

ISBN: 978-1-77469-477-0 (Hardcover)

ABOUT THE EDITOR

Miss Anita Laghulkar is presently serving as Assistant professor, Department of Food Chemistry and Nutrition, College of Food Technology Naigaon (Bz) Nanded, Maharashtra affiliated by Vasantrao Naik Marathwada Krishi Vidyapith, Parbhani. She obtained her B.Tech (Food Technology) in 2016 from College of Food Technology Naigaon & M.Tech (Food process Technology) in 2018 from Maharashtra Institute of Technology Aurangabad, Maharashtra. She started her career as Assistant professor in 2018. Her M.Tech specialization in reduction of acrylamide compound from processed Food. She has published 2 research paper in international journal. She has participated in many National and international symposiums and state seminar.

TABLE OF CONTENTS

LIST OF FIGURES

ABSTRACT

Food shelf-life can be described as the period required to experience general food quality and make it unsuitable for consumption. These undesirable qualities are defined by objective pointers, for example, the existence of off-scents and off-flavors, adjustments in texture attributes, and appearance alterations. These major changes can be subject to microbiological, chemical , or physical responses. Certain factors like temperature, pH, microbial and aw contamination, and cross-pollution during manufacture, storage, and commercialization can impact the product's shelf life. Regardless, shelf-life limitation affects socioeconomic factors, i.e. waste treatments and production costs, and risks related to food consumption. Meat and fishery items are unrefined substances prone to deterioration because of their chemical structure and increased water content. In these food items, the primary spoilage activities are because of microbial digestion and oxidative responses for the lipid elements. In new items, the shelf life is ensured by a fitting cold chain which may defer the emergence of microbial, enzymatic, or chemical reactions responsible for decay. This volume suggests that the impact of refrigeration temperature is typically reinforced through vacuum or modified atmospheric (MAP) packaging. In the last option, various gas applications can be utilized, regardless of whether the antimicrobial function is frequently controlled by CO2. Carbon dioxide concentrations in the environment greater than 20% can hinder or reduce the microbial development of aerobic microorganisms.

PREFACE

Food shelf life is the duration in which food retains an adequate safety standard and organoleptic perspective, this relies on four principal factors, specifically formulation, handling, packaging, and storage factors. Extraneous elements, for example, gaseous environment, storage temperature, and comparative humidity alongside internal properties of food decide the shelf capacity and perishability of food. Shelf-stable food sources have reduced water activity or potentially low pH, while perishable food varieties have inverse inherent properties. Stale food causes poor flavor, smell, gas accumulation, exudate discharges, and physio-chemical variations in appearance, color, and consistency.

Rotten meat or fish discharge vaporous amines like ammonia because of floral development or any compound product changes. Biogenic amines are likewise markers of food waste; for instance, putrescine, histamine, and tyramine usage have all been connected to hypertension and headache. Xanthines are constituent elements of purines and are applied for anticipating fish and meat shelf life. Xanthine oxidase formed to chitosan-ZnO NPs scattered on multiwalled carbon nanotube structures inside polyaniline grids was concentrated as a xanthine sensor. Expansion of ZnO NPs to carbon nanotubes creates better and quicker transportation of electrons between the strategic locales of biomolecules (ie, xanthine oxidase and anode with great chemical stability).

In processed food items, temperature (sterilization, chilling/freezing, pasteurization) is a critical component for delaying the commercialization period of food items. However, different obstacles can be considered for food preservation, for example, aw decrease (curing, salting, and drying), maturation, pH and redox potential management, MAP, utilization of additives, and the use of non-thermal treatments (i.e., ionizing radiations, HHP). New methodologies are being created to further develop meat and fish shelf life because of market and customer demands. Specifically, the principal drivers are financial costs (waste decrease) and interest in healthy and "natural" techniques. Among the arising techniques used to improve the shelf life of meat and fish products include, bioprotective treatment which is gaining immense popularity in the food business.

This volume mentions that bioprotective cultures comprise safe microbes (typically lactic acid microbes), which are purposely immunized to control the natural environments and reduce the intensification of spoilage and risky (pathogenic) microorganisms. The cultures can accomplish this goal due to their inclination to compete in the particular food grid and the creation of antimicrobial compounds, (for example, bacteriocins, volatile mixtures, and natural acids) that are obstacles for the undesired microorganisms.

Also, bioprotective cultures are chosen for limiting the effect of their development on the general organoleptic product qualities. Another emerging technique to delay shelf life depends on the utilization of substances gotten from plants and other natural elements, which can have antimicrobial and anticancer properties in food. This is, for example, found in essential oils and their constituents which are primarily terpenoids, phenylpropanoids, and terpenes. The inhibition properties of these molecules against a few microorganisms (including microbes) are well defined, regardless of whether data concerning the method of activity and the collaboration with other stabilizing elements are occasionally scant. Thus, more research is required to upgrade and normalize their utilization at industrial capacity. In any case, the antioxidant properties of these substances are because of how natural phenolic emerges from secondary plant metabolism.

CHAPTER 1

WHAT IS SHELF LIFE?

CONTENTS

The shelf life of a product is the period it may be stored without spoiling. Shelf life is always calculated based on the best-by date, but beyond that 'best-by' or 'sell by' is an arbitrary timeline set by the manufacturer, while 'use-by' is usually a federally regulated timeline and is usually directly related to food safety. It can also be identified as a shelf-life period. The shelf-life period is said to be the period from the date of manufacture until the date of expiration of use. After this term, a consumer may consider the product expired and must throw it away. In other words, shelf life refers to the period that may cause deterioration of the quality and safety attributes. It can also be described as the length of time before food deteriorates beyond the point of being edible. It is not just one number, but a range that you can use to work out how long certain food items will last.

Figure 1.1: Shelf life simply implies how long a product will take to be safe for consumption or use.

Shelf life is one of the biggest factors when storing foods long-term. It is important to understand what foods have the longest shelf life and why to correctly plan your food storage or prepping. More so, shelf life can also refer to the texture and other edible qualities when properly stored or to the length of time that foods may be stored before they lose these qualities. The shelf life of foods varies based on the product. Shelf life depends on various factors, such as expiration date, storage temperature and conditions, and special handling instructions. It is a way of ensuring that food is used while it is still fresh and delicious. The shelf life of a product is considered a crucial factor in determining its marketability, therefore resulting in lower sales (Zhukov et al., 2018).

Two of the major factors that affect shelf life are temperature and moisture (especially humidity). To understand shelf life, however, you must also consider product packaging, product quality at the time of purchase, storage conditions (cool, dry, and away from heat sources), and expiration date. For example, the shelf life of a drug depends upon factors such as purity, potency, physical stability, and chemical stability. The shelf life creates an optimal state for consumption. This means that it does not have any signs of deterioration that significantly decrease the edibility or aesthetic quality of the product. Moreover, it is also known as the "best before" date, which refers to the date on which a manufacturer guarantees that its product maintains the best quality.

One of the most common questions asked by consumers before they make a purchase is, "How long will the food or other product stay fresh?" It is important to note that shelf life differs from the expiration date, a more legal obligation to either the manufacturer or distributor of a food item. The shelf life of a food item is important to know (Weinbreck et al., 2010). A product that has been refrigerated for the majority of its life can sit out at room temperature for a certain amount of time before spoiling, depending on the type of product. Sliced meats can be kept for only a few hours, while cheeses and cured meats can be kept for months or even years.

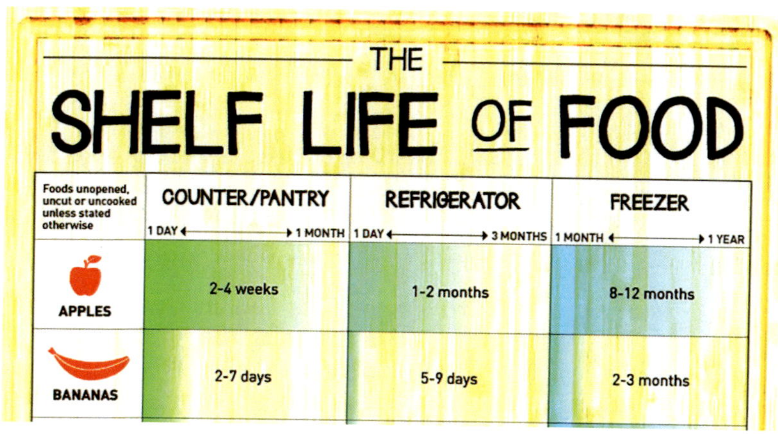

Figure 1.2: Self-life of Food.

It is important to note that shelf life goes beyond the theory of food products and other commodities stored on shelves at the local store. It applies further in the homes. Foods should be stored at room temperature (77F) or lower to maintain this period; otherwise, the food could easily pick up bacteria that could lead to illness. Food stored in the refrigerator should not be kept for over 2 weeks, and freezers must be used for longer storage (up to 6 months). This, however, is dependent on the type of food. Fruits and vegetables last longer than meats, which are already packaged and sealed for freshness (Villalobos-Delgado et al., 2019).

Food expiration dates have recently been drawing much attention. With claims that food is being unnecessarily thrown away, there are now some systems in place to help consumers make heads or tails of the confusion. One ought to stay away from artificial flavors, bacteria, stale food, and other components by learning what shelf life is and the factors affecting it. Having a good understanding of what shelf life is and how it is determined, you may be able to better predict when your food will expire.

Once a product has been opened and left unused for 5-7 days, the shelf life diminishes greatly. Food cups for instance must have a long shelf life or a short shelf life between heat treatments depending on the food product used. If a food has a long shelf life, it will probably have certain characteristics that prevent pathogen growth and spoilage. On the other hand, short shelf-life products do not require the same stringent processing conditions but may still have some magic flash or ingredient that helps preserve these drinks.

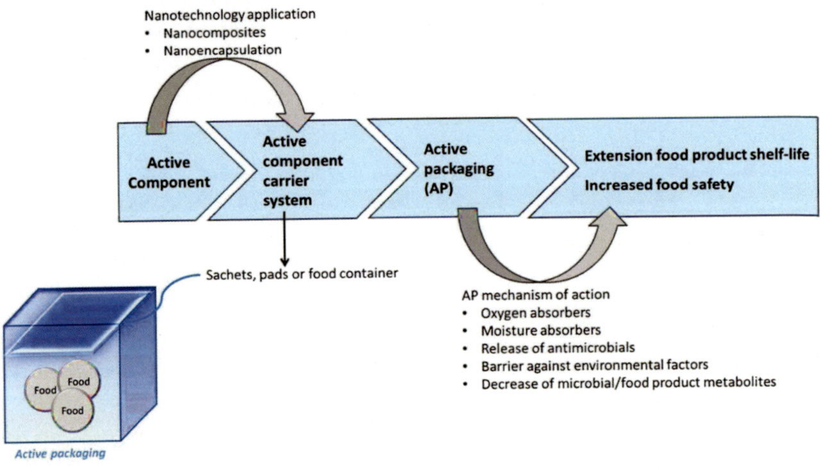

Figure 1.3: Shelf life has several factors to put up with (Source: Melini V, Melini F. *et, al.*,2021).

Indeed, Understanding the shelf life of your food is essential to food safety and efficiency in your kitchen. Keeping track of your food's expiration date is not only a convenience but also a way to ensure that you are serving the safest product available. When properly stored, many foods remain edible long after their expiration dates, some for several months or even years. Today, there is the existence of shelf-life tables which assures proper storage. Proper storage is crucial to food safety. Shelf life tables and charts are a numerical information source used in conjunction with allergen reactivity, diet history, food allergy, and intolerance data as well as diet recommendations to determine when food is unsafe for consumption.

Shelf life as a subject throws in the confrontation between processed and whole foods. High-quality foods can be a thing of the past. The quality of processed foods cannot match the quality of healthy and whole foods. Processed foods have little to no nutritional value and can never replace the benefits that whole foods can offer (Wrona & Nerín, 2020). Whole foods are packed with antioxidants and other compounds that play an important role in maintaining the health of your body and mind. It also confers to business. Shops are packed with junk food and treats designed to tempt your taste buds, but they're made of ingredients that will prompt an urge to make a mid-afternoon run for the snack machines at work.

Food freshness is a top priority for many people, especially as we become more health-conscious and focus on organic, whole foods. Fruits and vegetables are purchased at the grocery store and then placed back in the refrigerator but are usually never marked with an expiration date. The "sell by" date is just not specific enough to know when food needs to be consumed. It is important to know the shelf life of the food that you eat to maintain good health or avoid food-borne illness. Bacteria will grow increasingly quickly under certain conditions.

Until recently, many consumers didn't know the difference between a "use by" date and an "expiration" date, or the reason for the dates printed on the labels of their food products. Nowadays, consumers want to be informed, conscious and savvy. They want to know they are eating and buying safe foods that meet their needs (Winterhalder et al., 2015). While agencies such as FDA Food Code provides some guidelines, manufacturers still establish their shelf-life policy.

Companies need to determine at what point they feel the product is no longer in compliance with consumer expectations. This decision depends on a variety of factors such as product quality and its intended use. The best food companies perform shelf-life testing annually; for fresh products, it may happen every week. These companies have developed their internal methods of shelf-life determination.

1.1 WHERE DOES SHELF LIFE APPLY?

Shelf-life of a product depends on its chemical and physical stability. Products free from contamination will have a longer shelf-life than products in contact with air, light, extremes of temperature, etc. It applies after the expiration date on the package. The shelf life of a product varies depending on the nature and quality of the product. The shelf life of food items usually ranges from 30-90 days and that of cosmetics, shampoos, and body washes ranges from 0 to 60 days.

The Shelf-life varies for different fresh foods and can be affected by several factors including temperature, moisture, packaging, and pH level if the container used is airtight. For cosmetics, food and beverage products, medical equipment, and pharmaceutical treatments as well as tires, batteries, and many other perishable things, shelf life is an important consideration. Bread crumbs, white sugar, brown sugar, and flour are pantry items that do have a shelf life of 6 months, longer if kept in the refrigerator. Salt and baking powder can stay around for a year.

Pasta has a shelf life of 4 years. Ice can be stored for a year but need to be frozen. Hummus has a shelf life of a year or more. It also should be refrigerated. Cheese that is wrapped well has a shelf time of about three months. A head of lettuce will last for about one week in the refrigerator if it is not open yet (Wibowo et al., 2018). Snacks like crackers can last up to six months and cereals can last up to 10 years with proper sealing conditions.

Expiration dates found on a variety of food products can be confusing and obscure. While most food products do not require you to be super health-conscious, they can pose serious harm if eaten after the expiration date. Some products become hard, stale, contain harmful bacteria, or increase in fat content after some time.

1.2 SHELF LIFE BY THE CONCEPT OF EXPIRATION DATE

The concept of an expiration date, in which the quality of beer deteriorates with time, was born from the fact that beers from breweries near the sea were not very good in quality. This is why they had to be consumed quickly, i.e. before going bad. Expiration dates are supposedly put on foods to help consumers know if the food will still be good when they buy it. The truth is that expiration dates are only a suggestion for stores as to when the food might begin to decline in quality from the time it is processed. They do this to sell more food and make profits. Some stores and countries even have policies that force food makers to add expiration dates based on the time they opened their packages.

1.3 IMPLICATIONS ON SHELF LIFE IN THE COMING DAYS

There is optimism in the application of nanotechnology and data science to revolutionize the way people think about food expiration dates. Light technology uses a patent-pending process to eliminate bacteria from non-refrigerated items by superheating their environment (external pressure wave) into extreme heat which kills off any living things that may grow on them. The result is a local arctic-like environment that freezes walls of cans and boxes with brilliant cold light without freezing the contents inside (Uebele et al., 2013). There is also the removal of oxygen from industrial use in the packaging sector to preserve foods longer. Tracking of products is also on the rise.

Having a longer shelf life does not mean that a food item is safer to consume or more nutritious; it simply means it will last longer on the shelf surrounded by other products. We often take refrigerated and frozen foods for granted because we expect them to be fresh and safe to eat. However, many of these products will not stay fresh indefinitely, even when they are correctly stored. The problem is that during storage, potential hazards such as microbes (bacteria, molds, and yeasts), enzymes, toxins, and deterioration products from the foods themselves and packaging materials will develop. If these contaminants are present at high levels in food at the time of consumption, illness or even death may result.

CHAPTER 2

FOOD QUALITY

CONTENTS

2.1 INTRODUCTION

Figure 2.1: Food quality.

2.2 EVALUATING THE QUALITY OF FOOD

When people talk about "food quality," they usually mean everything a customer likes about it. Quality attributes include the following:

- It's not just about how big or small something is; it's also about how shiny and consistent it appears (including size, shape, color, gloss, and consistency) (Turner et al., 2020).
- Cosmetics
- Nutritional benefit.
- Manufacturing that is long-term and ethical.

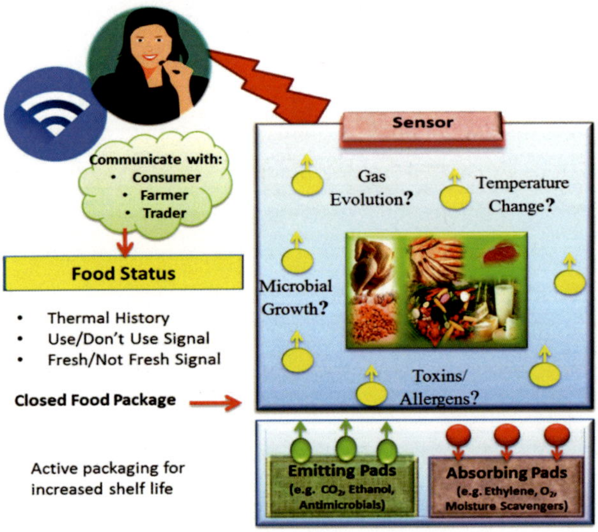

Figure 2.2: Food quality evaluation.

(Source: Mustafa F, Andreescu S. et al.,2018)

If food safety and legal standards are regarded as components of food quality, they assist customers in accepting the product and may be used as a marketing strategy to sell goods in countries with strict food safety regulations.

Food quality is a term that is frequently used, but it is difficult to define because what one person considers "acceptable" and "valuable" in one country may not be the same in another. As a result, depending on how it is defined, the term quality can be interpreted in a variety of ways. As a result, in the food industry, a food product's quality attributes or standards are frequently mentioned in the product specification. The quality of a product is determined by its adherence to these standards (Timakova et al., 2020). Quality management systems are used to ensure that food is produced by specifications and to remain competitive in markets that demand consistent quality and a predictable supply. Food safety and quality are important in the home kitchen, but they are even more important in large-scale food manufacturing and processing.

Traditionally, a large portion of food was prepared at home. There has been an explosion in the variety of processed foods, food for health / functional foods as a result of advances in technology and processing, as well as rising per capita earnings. An investigation into the food's safety is required. Food quality, both raw and processed, is a public health concern. As a result of the global and Indian food safety crises in the last decade, food quality and safety concerns have taken on a new level of importance. This is due to several factors, including people's lifestyles and eating habits change, more and more people are dining out. The handling of commercially produced foods involves a large number of people, which increases the risk of contamination. It's also worth noting that food is prepared ahead of time and can spoil if not properly stored. There are numerous pre-made and pre-packaged meal options available. Spices and condiments, as well as oilseeds, were traditionally prepared at home, and purity was not an issue. Spices, condiments, powders, and mixes are in high demand in today's cities and metropolises around the world. Raw food, in addition to processed meals, can pose a public health risk. The logistics of bulk food transportation are complex, and the time between production and consumption can be lengthy.

Risk assessment and safety management are critical to mass manufacturing and distribution operations (Sutton et al., 2016).

Because of changes in human sensitivity and the global spread of foodborne pathogens, the number of food-borne microbial infections is increasing. In the previous 25-30 years, the number of cases of food-borne illness increased by around 50%. A large number of people continue to be afflicted by unknown foodborne diseases. Because this is a global public health priority, active surveillance networks must be built at the local and global levels to find, diagnose, and recognize emerging illnesses. As a signatory to non-tariff trade agreements, India has assisted all nations in gaining greater access to global markets and opportunities for international business. Food safety and quality must be ensured in each country while also ensuring that imported commodities are safe to consume. Only strict food standards and quality control systems make it possible to maintain domestic food production while also facilitating international trade (Sewald & DeVries, 2003). All food producers must follow quality and safety standards, and their products must be tested regularly. Toxins accumulate as a result of air, water, and soil pollution, as well as pesticides used in agriculture. Food analysis is required to detect the presence of a wide range of components, both nutrients, and pollutants, due to the use of additives such as preservatives, colorants, flavoring agents, and stabilizers, as well as other chemicals such as stabilizers. The scope and career opportunities for this profession have grown dramatically as a result of the aforementioned reasons for a growing need for safe, healthy, and nutritious meals in a fast-paced food sector. Before looking at different career options, it's critical to figure out what you want to achieve. A solid understanding of the fundamental concepts of food quality, safety, and risk assessment would be beneficial in this course. Food standards and quality management systems would be beneficial as well.

2.3 FOOD QUALITY ATTRIBUTES

There are two food quality attributes which are: **Intrinsic and extrinsic attributes**

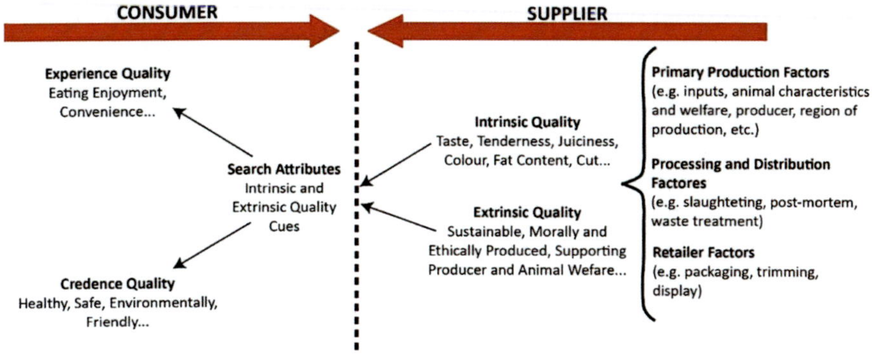

Figure 2.3: Intrinsic and extrinsic qualities.

(Source: Paiva T, Jacinto TA, Sarraguça MC, et, al., 2022)

A product's quality can be influenced by a variety of factors, including the technology used to create it. They are displayed in a quality triangle. Qualitative characteristics are formed by a customer's perception of a product's physical characteristics. A wide range of factors influences the perceived quality of agri-food products. Product dependability, safety, nutritional content, sensory characteristics, shelf life, and convenience (e.g., ready-to-eat meals) are all important factors in a customer's decision to purchase a product (correct weight, the right composition, etc.). Intrinsic features are those aspects of a product that are directly related to its physical properties (Sumner & Jenson, 2011).

External factors such as environmental impact and marketing influence can all have an impact on a company's ability to manufacture a product. Though not always related to physical characteristics, they can have an impact on whether or not a product is accepted by the target market. To name a few examples, the use of pesticides, antibiotics, and biotechnologies to improve product attributes can all have a significant impact on consumer acceptance of food products.

Extrinsic characteristics are mostly determined by marketing operations, but intrinsic characteristics reveal the essence of something. Appearance, color, form, and texture are all examples of typical inherent characteristics in this classification. Price, brand name, packaging, labeling, and product information are all examples of extrinsic attributes. Traditional food variables such as product characteristics and environmental impact are incorporated into our approach, which is typically associated with marketing considerations rather than extrinsic characteristics (Amani & Gadde, 2015).

According to Van Trijp and Steenkamp's theory, quality perception occurs at two points: when a product is purchased, the buyer has a quality expectation, and when the product is consumed, the consumer experiences the actual quality. When it comes to repeat business, customer satisfaction is more important than customer quality expectations. An extrinsic characteristic is not inherent in a product, such as a price, brand, or store name. As a result, these signals influence the customer's quality expectations. In most cases, internal cues are more important than external cues. Even if the use of pesticides has no direct impact on product features, it can alter consumer expectations about quality. According to a case study of the biological meat production chain, the quality of the meat produced this way exceeded even the most optimistic consumer expectations. The negative evaluation of the production system's appearance obscured the positive effects of the system's characteristics.

2.4 INTRINSIC QUALITY ATTRIBUTES

In terms of internal and exterior quality attributes, there are various categories. The intrinsic features of a product include its safety and health considerations, shelf life and sensory properties, convenience, and product reliability.

2.4.1 Sensory Characteristics and Shelf Life

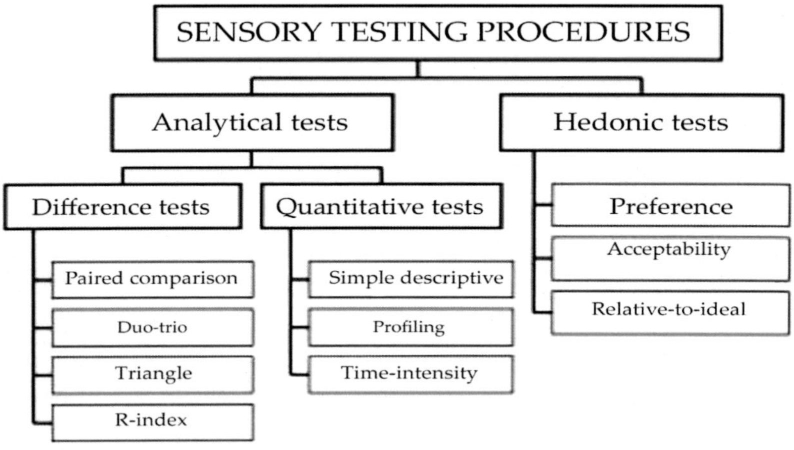

Figure 2.4: Sensory characteristics and shelf life.

The sensory qualities of a product are defined by its physical attributes and chemical makeup. Agrifood items are often perishable. After a fresh product is harvested or processed, the food degradation process begins, which has a detrimental influence on the sensory qualities. Processing and/or packaging are meant to postpone, stop, or minimize degradation processes to extend shelf life. A product's shelf life may be regarded as the time between harvesting, processing, and packaging the product and the point at which it becomes unfit for eating. Microbiological, (bio) chemical, physiological, and/or physical characteristics can all impact shelf life. Freshly picked peas, for example, are ruined within 12 hours, but canned peas may be stored at ambient temperature for two years (Angourakis et al., 2015). Unacceptability is generally characterized by shifting sensory properties, such as the formation of rotting odors or a sour taste as a result of microbial decomposition. The real shelf life of a product is controlled by the pace of degradation processes; for example, even if a product is safe and not harmed by germs, it will become unattractive owing to its grey tint.

Microbiological activity in meals can cause food deterioration and the production of unwanted sensory qualities, such as texture loss and the development of off-flavors and off-colors. In many circumstances, the food incorporates infections and becomes harmful before any modifications in sensory aspects are observed.

Non-enzymatic browning and oxidative reactions, which generate changes in appearance and lower nutritional value, and oxidative reactions, including auto-oxidation of lipids, which vary the flavor, and bleaching of plant colors, are examples of chemical processes that may restrict product shelf life. Generally, chemical changes occur during agri-food processing and storage. The non-enzymatic browning process also adds to desired quality traits such as bread crust browning and fried meat browning.

Enzymes are generated throughout **biological processes** when the integrity of plant or animal tissue is disturbed. Cutting fresh vegetables, for example, triggers a variety of enzymatic activities, including browning by phenolases and off-flavor generation by lipoxygenase. Under comparable circumstances, biological processes, such as cabbage fermentation, are managed and utilized to generate more digestible food. One noteworthy example is beef maturation by increasing temperatures, whereas germ surface multiplication is regulated by ultraviolet light (Calligaris et al., 2016).

Physical changes are commonly connected with agrifood mismanagement throughout harvesting, processing, and distribution. Changing temperature and humidity conditions during storage and transit may induce desiccation of humid materials, swelling of dry items, or phase changes. Other prominent physical processes that result in unfavorable product qualities include emulsion breakdown and phase separation. Physiological reactions in fruits and vegetables typically occur during post-harvest storage and are highly dependent on storage circumstances. The items continue to respire and ethylene is created, which has a substantial influence on usual post-harvest concerns.

When a product is subjected to a single extreme response, several variables might contribute to its short shelf life. Lipase activity, which results in short fatty acid chains, as well as fatty acid oxidation, can both contribute to rancidity. To solve the issue of low quality, it is necessary to identify the underlying mechanism. When it comes to shelf life, the quickest response is frequently to blame. Changes in the texture of bread, for example, are frequent before mold growth. Physical deterioration occurs quicker than microbiological changes in this situation, and degradation is the primary predictor of shelf life. It is standard practice to suppress, reduce, or avoid the principal cause of a component's shelf life limitation to extend that component's shelf life. Long-term storage might indicate more progressive degeneration (Sarkar et al., 2022). Frozen foods, for example, have a longer microbial-free shelf life, but chemical and physical interactions cause color and texture changes after one to twelve years. Understanding the many mechanisms that determine shelf life and sensory qualities is critical for preserving product quality.

2.5 CONCERNS ABOUT PRODUCT SAFETY AND HEALTH

Product safety and health concerns are important aspects of intrinsic quality. Diet and dietary composition are important factors of health, with nutritional imbalances harming human health. To meet these nutritional goals, the food industry is progressively manufacturing functional meals. Low-fat and low-cholesterol diets, as well as those supplemented with vitamins and minerals, have been shown to improve human health. Food safety refers to the need that foods to be safe while yet posing a reasonable risk. A hazard is a potentially detrimental source of danger, whereas a risk is an assessment of the likelihood and severity of human health injury. Toxic substances, pathogenic bacteria, physical forces, and natural disasters are just a few of the many causes that might jeopardize food safety (Roccato et al., 2017). Negative health implications can appear across a range of periods, some of which are brief (such as allergic responses or food poisoning), while others are long-term and chronic (such as cancer, heart disease, and vascular disease) (such as cancer, heart disease, and vascular disease). These long-term impacts might be the result of nutritional imbalances or long-term exposure to dangerous substances.

Pathogens are microorganisms that cause disease. Mold and bacteria are both pathogens that can cause illness. It is critical to distinguish between food contamination and food poisoning. Food contamination occurs when live bacteria are discovered in food and then enter the human body, causing disease. Salmonella, Shigella, and various types of *Escherichia coli*, as well as *Campylobacter jejuni*, are the most common bacteria that cause food poisoning. Salmonellosis can be found in beef, turkey, chicken, eggs, pig, and raw milk products. Poor personal hygiene is a common source of food-borne Shigellosis, which can be spread by eating raw or undercooked fish and poultry, salads, fruits, and vegetables. Certain strains of *Escherichia coli* are commonly responsible for traveler's diarrhea. *Listeria monocytogenes* is another pathogen found in food.

Listeria species may breed in sludge, sewage, animal waste, and dirt, among other places. Listeria is present in a variety of dairy products, including raw milk and soft cheeses. The contamination of various commodities is assumed to be the result of zoonotic transmissions or the transfer of sickness from animals to humans. *Campylobacter jejuni*, the most prevalent cause of acute bacterial dysentery in humans, is an environmental bacterium, however, the germs are passed through animal digestion. *C. jejuni* has been detected in the feces of numerous main meat species, including chicken. Pathogenic bacteria (enterotoxins) and molds (mycotoxins) in food produce dangerous substances, which cause food poisoning. Chemicals are present in both raw and processed foods (Pratiwi et al., 2015). Toxic food can produce a wide range of symptoms, ranging from short-term issues like nausea and vomiting to long-term issues like cancer or abnormal liver histology.

2.6 TOXIC CHEMICALS ARE CLASSIFIED AS DANGEROUS

Toxic substances can contaminate the food chain in a variety of ways. Poisons can form during storage and processing, while others may exist naturally in raw materials. Human health is jeopardized by agricultural pesticide residue, veterinary medications, and disinfectants. To make matters worse, food and color additives are known to be especially hazardous. These are the criteria for evaluating toxicants that represent a danger to food safety. Chemicals, both natural and man-made, can be found in the environment. Hydrazine, a toxin found in raw fresh mushrooms, is worsened by damp and humid storage, for example. Fish and meat that has been grilled, smoked, or deep-fried, for example, may contain heterocyclic amines.

2.7 EXTRINSIC QUALITY FOOD ATTRIBUTES

Extrinsic quality characteristics include those of the manufacturing process, the environmental impact of the food item, and manufacturing and marketing issues. While extrinsic quality features do not always have a direct impact on a product's physical qualities, they can have an impact on a consumer's perception of quality (Castrica et al., 2021). Marketing efforts, for example, may influence client expectations even though they have little to do with the actual product.

2.7.1 Characteristics of the Production System

The manufacturing process of a food product is referred to as its production features. Pesticide use in fruit and vegetable agriculture, animal welfare in cattle breeding, genetic engineering to improve product quality, and the use of specialized food preservation techniques are just a few examples. The effect of production system elements reduces the acceptability of a product. Concerns have been raised about the public's acceptance of new genetically engineered foods, as well as genetic engineering in general.

2.7.2 Concerns About The Environment

Figure 2.5: Concerns about the environment.

The bulk of agricultural commodity-related environmental issues is related to packaging and food waste management. While flavor and nutritional value are intrinsic quality elements related to human aims, environmental quality components of food may be linked to a broader social purpose. Consumers want to buy environmentally friendly meals for several reasons, including their own and the environment's health. To decrease the environmental effect of packaging waste, European legislation has been implemented (O'Brien & Bentley, 2015). Since 1997, food packaging manufacturers have been required by law to assist with material recycling and thereby reduce packaging waste. In terms of waste management, inefficient processing poses a greater problem to processors than to consumer quality.

2.8 PUBLIC RELATIONS AND MARKETING

Marketing has a wide range of effects on product quality. Marketing actions (branding, pricing, and labeling) shape consumer expectations for quality, establishing extrinsic quality attributes. Marketing can alter the quality of an experience by modifying its credibility (which customers can judge) (which can be checked by customers themselves).

2.8.1 Food Safety; Quality Management Systems

Figure 2.6: Food safety; Quality Management Systems.

Quality management systems (QMS) are created to ensure the product's quality and safety. A quality management system (QMS) enables the many components of a company to function smoothly and efficiently. A system gives a company a competitive advantage by increasing its marketing and sales choices, allowing it to attract new customers while keeping old ones. Communication, morale, and job satisfaction are improved by including the entire team in a quality management system. ISO 9001-2000 is a collection of rules for quality management systems. ISO 9001 is a set of six clauses that act as a guideline for a company's quality management system (Nicoli, 2012). It is not prescriptive, but it enables managers to design and implement a quality management system that is tailored to the specific needs of their firm.

2.8.2 Food Quality Management Systems

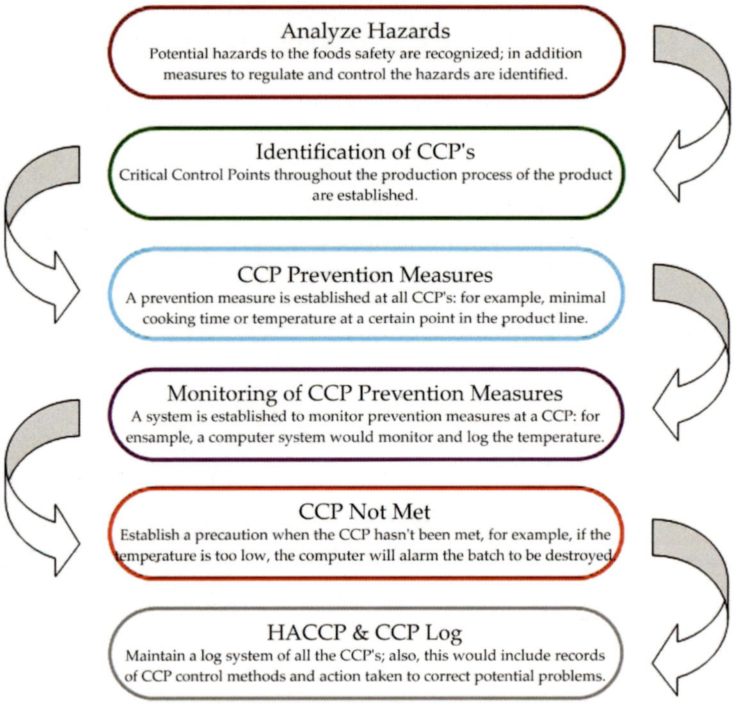

Figure 2.7: Food quality management systems.

Food quality management has grown in popularity over the last decade as an approach for ensuring food health and safety. This is the result of a slew of issues, including shifting consumer preferences, increased competition, environmental concerns, and government interests. As a result, the food market and agri-food supply chain are being severely disrupted. The complexity of food and food components, which include unpredictability, a short shelf life, possible safety problems, and a wide spectrum of chemical, biological, physical, and microbial activity, exacerbates the problem. To address this issue, food quality management systems must be improved regularly, and staff members must be aware of new technology and management practices (Chang et al., 2022). Furthermore, because it is unpredictable and constantly changing, human handling is essential for quality control. As a result, the outcome of agriculture and the food industry as a result of the collective effort of humans seeking quality is far more unexpected than widely assumed. HACCP, ISO, and BRC (British Retail Consortium) are only a few of the quality assurance systems used in the food industry. To assure food safety, various systems, as well as combinations of these systems, are utilized.

The quality management system (QMS) is a structured system that defines the organizational structure, responsibilities, processes, procedures, and resources used by the food industry to coordinate and regulate commodity manufacturing. It is a preventative strategy that identifies problems before they occur by performing continuous process and product evaluation, auditing, and fault correction to preserve contractual product quality while increasing production effectiveness and efficiency. Three key food quality approaches are widely used to ensure product conformity, meet consumer expectations, and aid in due diligence efforts:

Good Agricultural Practices (GAP):

GAPs are basic food safety principles that aim to limit biological, chemical, and physical hazards across the food supply chain.GAP practices include site selection, land use, wildlife and habitat protection, water, fertilizers, pesticides, GMOs, integrated crop management (ICM), animal feeding techniques, worker hygiene, field and facility cleaning, cooling, and transportation.

Good Manufacturing Practices (GMP)

All individuals who come into direct contact with food, as well as those who work on surfaces that may come into contact with food and those who work with food packaging materials, are required by GMP to follow all necessary sanitation and hygiene procedures to protect food from contamination from both direct and indirect sources. GMPs are a requirement for HACCP and are frequently included in the applicable legislation. GMPs are the basic minimum of hygiene requirements that must be met for a company's goods to be safe and consistently of high quality. All parties participating in the animal feed chain, including those involved in the manufacture, processing, storage, transportation, and distribution of feed and feed components, must comply (Dawson et al., 2018). Each level in the chain is accountable for the actions that fall under its purview.

Some examples include food establishment design and layout, structural condition and maintenance of food establishments and equipment, provision of adequate plant equipment and facilities (e.g., toilets, drainage, ventilation), personnel hygiene and food handler training, cleaning and sanitation, pest control, storage, distribution, and transportation, and waste management.

2.9 HAZARD ANALYSIS AND CRITICAL CONTROL POINTS (HACCP).

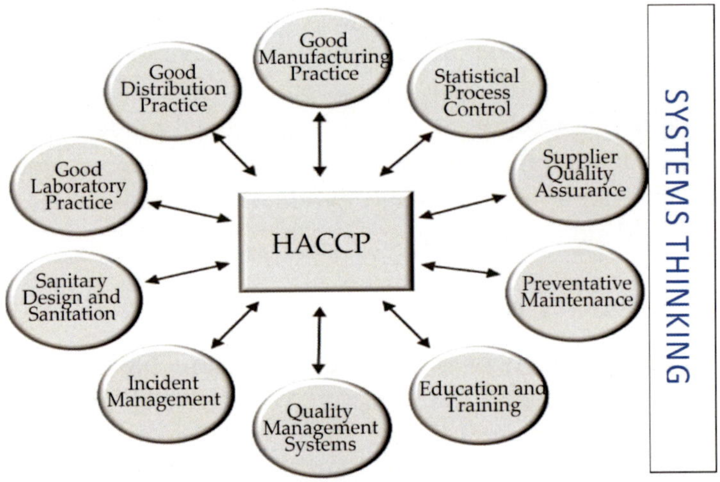

Figure 2.8: Hazard Analysis and Critical Control Points (HACCP).

(Source: Lee JC, Daraba A., et, al.,2021)

HACCP is a well-known risk management technique that assists feed and food producers in identifying critical control points for contamination caused by microbiological (e.g., salmonella), chemical (e.g., pesticides), physical (e.g., glass), or allergic (e.g., nuts) sources. Rather than relying on normal inspection and quality control approaches to indicate compliance or failure, HACCP conducts a proactive and systematic risk assessment, choosing appropriate control and monitoring systems, particularly those deemed critical to the safety of the product.

Minimum HACCP and GMP requirements are established by European legislation (Nirupama et al., 2010). It does not, however, indicate how these criteria should be implemented in reality. As a result, several businesses have formed their quality control standards, while others rely on governmental and private food safety and quality control standards.

2.9.1 Application Fields for HACCP

To satisfy public health and hygiene regulations, food enterprises must install hazard control and prevention systems (HACCP) or components thereof employing prescribed methods. The creator and the supplier are both responsible for food safety. Food makers are responsible for assuring the safety, nutritional value, and acceptance of their goods. As a result, the manufacturer must take into account the Union's standards and follow Community practices. In EU member states, the HACCP system is used to detect food safety threats and the origins of quality problems in the food sector and distribution.

2.9.2 Result of HACCP Application

In recent decades, food-related diseases, intoxications, and adulterations have become more frequent around the world. Previously, guaranteeing food safety was a governmental duty that was administered by a state-appointed regulatory authority. On-site inspections and sampling-based laboratory testing were critical components of the quality control system. Activities conducted in reaction to an issue, on the other hand, did not affect the security of the goods listed below. This endeavor was unsuccessful, prompting EU and developed-country experts to evaluate and revise food safety legislation and systems.

Instead of certifying the completed product, a better process monitoring and control system was created. While the producer and distributor have defined duties in the food processing industry, quality assurance methods are an important part of the process. Even though the ISO 9000 standard series applies to all aspects of economic life – establishing a systematic approach through process control and clarification, task assignment, and the establishment of unambiguous responsibility relationships – it doesn't address food safety concerns. As a result, a food-safety-specific quality assurance system that is applicable throughout the whole food chain and supports both official control and international commerce must be built (Nagi et al., 2012). The HACCP system handles these circumstances. HACCP integrates both Good Hygiene Practices (GHP) and Good Manufacturing Practices (GMP). The GHP and GMP guidelines, HACCP, and the ISO 9000 quality assurance component are all built on top of each other, which is a system advantage. The favorable benefits and advantages of any pyramid system can be combined.

2.10 FOOD QUALITY STANDARDS

Figure 2.9: Food quality standards.

The WTO agreement on technical trade barriers (TBTs) defines "standard" as follows:

A document that has been authorized by a recognized entity that contains optional criteria or qualities for things, associated procedures, and manufacturing methodologies. Customer or third-party standards serve as a framework for food companies to develop policies and procedures, as well as to implement a quality management system that meets stringent standards for food safety, food quality, and customer requirements, as well as for the protection of people, animals, and the environment (Maertens et al., 2012).

Several reasons have contributed to an increase in public and private food safety and quality standards, including expanding customer and consumer demands, greater interest in food safety, improved adherence to legal and institutional requirements, and increased efforts to acquire a competitive edge. A thorough customer- or third-party-developed standard will contain the following:

- Employee training
- Objectivity
- A step-by-step procedure manual
- Standardization of procedures
- Important aspects of control
- Sample and analysis itineraries
- Reporting systems
- Evaluation protocols

Not only must these process and procedural controls be developed, but there must also be a method in place to validate the efficacy of a food company's quality management system.

2.11 AUDITING

Figure 2.10: Auditing.

Audits are often performed to ensure that a food sector complies with applicable rules and internal standards. Auditing comprises reviewing documents, doing inspections, and conducting interviews to ensure that a standard is followed. The three categories of audits are first-, second-, and third-party audits:

- Internal audits, also known as first-party audits, are performed within the business by internal auditors. They are used for self-evaluation, business standardization, and continuous improvement (De Corato, 2020).
- A principal organization conducts a third-party audit of one of its suppliers or subcontractors to determine the effectiveness of the

supplier's quality system in supplying safe materials or products. If a supplier refuses to assist with the audit, they risk losing a customer.

- A third-party audit is performed on a large organization by an external and independent authority (known as a certification body), which uses its skills and abilities to audit the feed or food firm and verify compliance with a set standard. This independent audit and certification of the quality management system by a third party are used to demonstrate due diligence and/or as a marketing tool.

Organizations' and Corporations' Policies and Strategies to Promote Food Quality

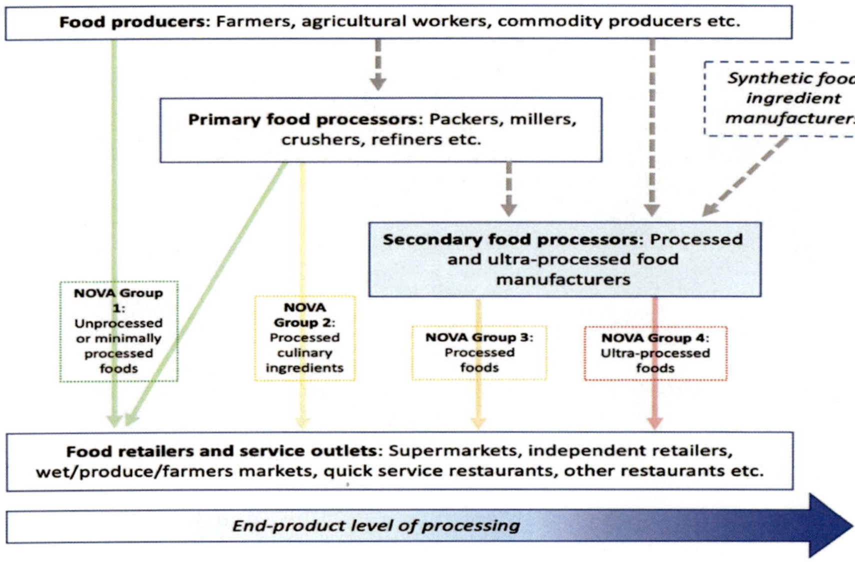

Figure 2.11: Organizations' and corporations' policies and strategies to promote food quality.

Improving goods and processes is critical for developing competitive demands for innovation and cost savings in the food and agriculture sectors. In this industry, the three pillars of success are planning, control, and improvement.

The term "**quality planning**" refers to the ongoing process of producing goods or providing services that meet the needs of customers.

Quality control refers to the techniques used to achieve the goals, such as comparing actual quality performance to quality targets and taking corrective action if there are discrepancies.

Quality improvement is motivated by a desire for change. Businesses must learn to facilitate, communicate, construct procedures, and reward systems in response to changes in the internal and external business environment (Derkach et al., 2020).

Quality enhancement necessitates a collaborative approach including both management and employees. Because change disrupts established systems, management should be the driving force behind it. As a result of their routines, many firms have inflexible organizational structures. They stifle growth and will only be tolerated if there is a sufficient enough degree of devotion. Quality improvement refers to a systematic approach to system improvement. This method includes documenting, measuring, and analyzing. Quality improvement goals include increasing customer satisfaction, raising quality standards, cutting costs, increasing productivity, and speeding up the process.

2.12 CONCEPTS AND APPROACHES FOR PROCESS IMPROVEMENT

The following are three key phases in the process of system process improvement:

Make a flow chart that depicts the operation.

- Gather process information, identify each stage, and determine the inputs and outputs, as well as the people involved and the decisions that must be made.
- As needed, keep track of time, cost, working conditions, employee situation, environment and waste, accidents and/or safety hazards, revenues and/or profits, quality, and customer satisfaction (Manzocco, 2016).
- Create a flow chart that accurately depicts the process, including key tasks and choices.

Investigate the technique in depth.

- Inquire about the procedure: does it make sense? Are there any stages or activities that are missing? Are there any procedures or acts that are repeated?
- Inquire about each step: is it necessary, may it be eliminated, adds value, and is there any waste at this point?
- Look for opportunities for process improvement by asking questions like, "What causes recognized issues, and how may they be eliminated?" Is it possible to enhance the procedure by shortening it or decreasing its cost? Is it possible to speed up the process by providing a higher standard of living?

Proceed with care.

Depending on the results of your analysis, redesign the approach. Do papers improve through time, just as projected metrics like time, space, waste reduction, employee turnover, accidents, safety concerns, improved working conditions, revenue/profitability, and customer happiness do?

In the 1970s, the industrial industry introduced TPM (Total Productive Maintenance). TPM is a mechanism for determining and notifying the cause of equipment failures and system downtime. Rather than viewing maintenance as an unavoidable evil, TPM set the lofty goal of zero breakdowns. Hammer and Champy used the term "business reengineering" in 1993 to describe a new approach to improvement (Manzocco et al., 2010). Re-engineering (also known as process redesign) can increase work quality and speed while cutting costs by altering core procedures. Reengineering extends beyond procedures to include the full value chain, as well as fundamental redesign and restart. It is often used when the amount of improvement required exceeds what can be accomplished by minor operational changes. Deming developed a structure (the Deming cycle - Plan-Do-Check-Act) within which all stakeholders may continuously discuss and suggest improvements.

To organize improvement actions, the PDCA cycle is employed.

Plan

Conduct an examination and documentation of current processes. Collect data and analyze it to detect potential issues. Compile data and develop a strategy for improvement. Indicate which metrics will be used to assess the strategy.

Do

First, test the concept on a small scale. Methodically collect and organize data for evaluation.

Check

Analyze the information acquired during the "do" phase. Compare the results to the initial aims of the planning phase (Martins et al., 2008).

Act

Once the results have been validated, standardize the new technique and make it available to all process participants. Carry out the training for the new way. If the results are not as expected, either adjust your strategy and retry, or finish the work.

2.13 IMPROVEMENT OF QUALITY IN THE FOOD INDUSTRY

The food sector is still attempting to better the quality of its goods. The efficiency of Total Productive Maintenance (TPM) is often the primary focus of success stories in the food sector . To understand why food improvement initiatives are so rare in today's food business, examine the following factors:

Integration into problem-solving activities is frequently made more difficult by language difficulties when operators or staff lack education (noise, unpleasant smell, and excessive humidity) (noise, bad smell, and high humidity). In most circumstances, professionals in the appropriate departments are responsible for identifying who is accountable for revisions in the data (Kuijt, 2009).

It's tricky to upgrade equipment that isn't widely available. The information is shared with the device's maker, who also conducts adjustments. User preferences and requirements are often not communicated, and consequently, the equipment options accessible from suppliers are constrained. A lack of adequate information on high-level performance results. Operators at the lower levels, in particular, are not sufficiently briefed on quality results.

In the majority of circumstances, improvement actions go unrewarded.

In the food sector, soft organizational tactics are unique.

In general, the culture is more proactive and solution-oriented, with less time spent on dispute and analysis. Techno-managerial strategies for quality improvement urge an emphasis on crucial competencies, which must be reinforced by relevant management activities. Management improvement demands the use of proper measures and information systems. This document should combine crucial quality and safety elements associated with the underlying capabilities of technology (Fan et al., 2009). Operator and employee training should focus on attaining the technological and administrative competence necessary for the quality control points for which they are accountable. Another key management job is team development, which cannot be forced and requires excellent leadership talents. Managers should not only construct teams; rather, they should facilitate team creation in a way that allows for organic growth.

2.14 MARKET FAILURES AND PUBLIC ACTIONS FOR FOOD QUALITY

In a pluralist, liberal market democracy, state measures may be justified when instances of market failure, imperfection, or non-existence are acknowledged or anticipated. In other words, in the absence of such failures, we would believe that private interactions are sufficient to resolve the issue at hand and that no government regulation or involvement is required. Food safety will be regarded as a symptom of widespread market failure from now on, and regulation will be the norm. It is suggested that in the case of food quality, market failure is unlikely to occur and that private market or commercial behavior should be expected to prevail, though the public sector must still play a critical role in ensuring the necessary infrastructure and regulatory apparatus are in place to allow for product differentiation and quality.

The consequences of food safety violations are frequently delayed, dispersed geographically, and difficult to predict. When an issue is separated from its origins in time and place, it frequently manifests itself in this manner. Food poisoning symptoms appear hours, days, or even weeks later depending on the structure of the food chain if a meat processor fails to maintain microbiological cleanliness. In one country, a failure may appear to be a success in another. This can make tracing the source of the problem extremely difficult. Under these conditions, even well-known food-chain enterprises would not consider self-imposed health and sanitary criteria to be adequate protection (Galić et al., 2009). A system of public regulatory control is required due to the specific circumstances that allow the unscrupulous to take advantage of shortcuts. Food safety may be viewed as reflecting market failure components on a less practical and more philosophical level. When food components are free of viruses and toxins, all consumers benefit from the lack of contamination. Each consumer's satisfaction with being "sickness-free" has no bearing on the happiness of others. It is extremely difficult, if not impossible, to exclude non-payers (non-consumers) from this liberty.

As a result, one could argue that safe food practices imply nonrivalry in consumption and non-excludability, both of which are characteristics of market failure. As a result, we can anticipate a lack of food safety on the market. Having said that, and in contrast to other well-publicized examples of market failure, there is no reason to underestimate the significance of food safety. Indeed, because we all consume food regularly and the costs of over-provisioning for food safety are widely disseminated and shared, there is a tendency to overestimate the need for safety. Indeed, when we examine the current state of full disclosure and the blame culture that surrounds every food safety failure, we discover that the government has a strong incentive to exercise extreme caution. Again, the government anticipates that such constraints will be widely accepted due to their low spread-out costs. As a result, regulators will scrutinize food safety with vigor. This is accomplished by imposing cleanliness, purity, and hygiene criteria throughout the food chain, which are supported by monitoring, sanctions for violations, and a responsibility structure based on an assumed duty of care. The licensing of processes, facilities, and activities imposes many of these constraints.

To determine the appropriate administrative level for enforcing these limits, technical efficacy should be used. Because food production, catering, and retailing are decentralized, the majority of regulation and monitoring is done by local governments, either through national law sponsored by Member States (MS) or through regionally decentralized governance institutions. There will undoubtedly be differences between EU member states in terms of which ministries are responsible for these criteria, such as health, agriculture, the environment, or consumer affairs. In response to a decade of well-documented food safety violations, several countries established Food Standards or Food Safety Authorities. As the EU's single market expands, these challenges will resurface (Kendall & Dimond, 2007). Because of the free movement of commodities on the market, the harmonization of food safety norms and laws, as well as mutual recognition of Member States' processes and harmonized consumer information and labeling, has increased. When food safety fails, it is thought that isolating the offending country or area is the best course of action. As a result, free trade within the EU is temporarily suspended regularly. The commercial component of food safety is currently being investigated as a source of authority. In theory, it is at the EU level, as with other trade difficulties; however, as evidenced by the lengthy French embargo on UK beef cleared by the EU veterinary authority, this procedure is not widely recognized.

Furthermore, the EU established a European Food Safety Ability to provide scientific and technical assistance, collect and evaluate data, identify and report problems, and communicate with the public about any food safety concerns. Food safety is primarily a concern for the food chain, rather than agriculture. The majority of critical food safety risks are expected to manifest themselves in agricultural inputs produced upstream of farms or in post-farm transformations. This emphasizes the critical importance of product traceability throughout the supply chain. This, in turn, necessitates the use of globally recognized methods and labeling systems. These must be negotiated on a national scale and coordinated on a supranational and global scale. Food safety and quality may be jeopardized if traceability and the means of collecting it clash. After determining the requirements for tracking food components for food safety, the same methods and procedures for tracking food quality can be implemented (Howey, 2015). As a result, quality assurance systems bear some of the burdens of ensuring food safety. Customers may be misled as a result of this. Organic food is a great example of this. Organic agricultural licensing and certification systems are well-established, and the organic marquee or label is well-known.

However, it is a food quality indicator rather than a food safety signal. There is no scientific evidence to suggest that organic food is any safer than conventional (or inorganic) food. If this statement were incorrect, it would imply that eating non-organic food harmful to one's health (albeit a minor one) (although a slight). This would imply that traditional food production systems that rely on inorganic fertilizers, crop protection agents, and animal health treatments tolerated and even pushed by member states have accepted some public health mistakes. If this were the case, no Minister or Commissioner in charge of food safety would be able to sleep soundly at night. As a result, food safety regulation entails a complex licensing and control system. Crop protection, veterinary, and animal health product approval and registration are the first steps. It allows processes, facilities, and personnel from farmers to downstream, including food processing and manufacturing at all stages; transportation; storage; distribution; retail and catering; and import and export ports. It addresses occupational health and safety as well as family counseling on these topics. It also incorporates integrated large-scale public information systems to aid in the explanation of existing policies, as well as information on commodities and home management, family nutrition, and health principles.

Finally, there is a significant area of market failure in food safety, which has resulted in the formation of a complex regulatory regime at the EU, MS, regional, and municipal levels. However, this is not the same as agricultural policy and should not be confused with it. The CAP should be compatible with a food safety policy, and some of its financial sources may help to achieve the food safety law's objectives. Regardless of whether the EU has an agriculture strategy, food must be safe. Food quality is not, in general, the result of market failure. It is difficult to define what constitutes exceptional brilliance. Commercialization and marketing are used to aid in the process of product differentiation, which is critical for achieving a high level of quality. This is achieved through proper packaging and display of the goods. The plan is to divide the market into segments. Recognizing that some consumers have the financial means and desire to buy (typically in smaller quantities) at a higher price than the average allows businesses to provide a better value product and, depending on the costs associated with delivering the higher quality, generate higher returns. Numerous quality qualities are produced by plant or animal species or the manufacturing process, and as a result, they cannot be examined or evaluated at the point of consumption. Organic foods, pastured eggs, Charollais cattle, and high-quality vintage wines are all examples of this. There is no way to verify that the finished product is what it claims to be (Hedegaard & Skibsted, 2013). Consumers trust the traceability system because it ensures product quality. This is primarily based on business practices and firm reputations for quality, both of which would suffer if they were discovered to be cheating. According to this theory, self-interest is the most effective guarantor of honesty.

Private sector initiative, ingenuity, and flair are required for success at any level. High-quality merchandise necessitates meticulous attention to detail, as well as brand development and marketing. These treatments are costly and, in some cases, harmful. Whether they are recipes or modified versions of ready-to-eat convenience meals, many new food items fail. As a result, the private sector is primarily responsible for the production and promotion of high-quality goods. Despite the popular belief that the production and distribution of high-quality food products must be handled entirely by private enterprises, there are several compelling reasons for government involvement in developing and maintaining quality. The government has traditionally played two roles in the development and diversification of quality marks: encouraging companies to innovate and protecting consumers. The first function can be explained by the fact that small, geographically dispersed businesses lack the necessary skills, drive, expertise, and resources to develop new goods and processes (Hosseini & Jafari, 2020). These structural barriers necessitate government intervention to facilitate the formation of producer-owned enterprises capable of carrying out these quality-improvement operations. Second, for consumer protection, it is necessary to establish mechanisms for monitoring the companies that develop and operate assurance schemes, as well as to conduct periodic audits to ensure that the traceability systems are operating effectively and are free of fraud. As a result, food safety public policy serves the following purposes. To counteract the CAP's discouraging influence. Four decades of government-mandated intervention 'quality' requirements under the Common Agricultural Policy have obligated producers of a wide range of major commodities to produce to intervention 'quality' standards rather than market standards. This category includes grains and oilseeds, as well as meats and dairy products. It encouraged standardization and a commodity mentality, which led to a lack of interest in product differentiation and market demands.

Marketers argue that the CAP should be the last device used to market alternative concepts because it is the source of the problems in the first place! In contrast, the Rural Development Regulation (1257/99) item on processing and marketing allows money to be used to encourage and support the formation of schemes or farmer's organizations devoted to improving quality and marketing. This assistance can assist with capital investment funding, but the vast majority of such investment should be funded privately rather than through government subsidies. People need to be educated. There are obvious disparities in access to information. The first issue is the current information imbalance. In terms of manufacturing conditions and true quality, the seller has a significant advantage over the consumer. As a result of this massive public effort, nutrition labels, use-by dates on perishables, and information harmonization in the single market – the use of symbols to help overcome the occurrence of many languages – have all emerged. contribute to the development of world-class infrastructure. Each line must be slower than a commodity line to achieve the quality that luxury products demand. This may result in an increase in costs, at least at first. A growing number of people believe that commodities labeled "local and regionally produced," such as rare breeds or foods prepared in traditional methods, have inherent value and can benefit the communities that traditionally produced them (He et al., 2018). Co-ops and other producer or community groups, for example, can help coordinate and facilitate the processing, shipping, and sale of these items. As a result, the government may need to provide funds to support and coordinate initiatives. There are several paid options available, including the formation of co-operatives (producer groups), information, support, and training. Please keep in mind that this list does not include yearly financial assistance.

To counter the dominance of highly concentrated food producers or merchants, various community approaches can be used. National (or EU) chain stores may thus be required to keep a certain amount of shelf space available for locally or regionally produced items. Public funds can be used to establish farmers' markets or to fund organizations that promote farmers' markets and agricultural commerce. So that animal processing capacity can be calculated and scaled appropriately. In slaughterhouses, this is especially true. These businesses operate on a massive private scale. The removal of known hazard material from BSE-affected animals, for example, necessitates the supervision of qualified veterinary inspectors. Small abattoirs may be forced to close if they are compelled to pay for the time used by these meat inspection services. Because there are fewer slaughterhouses, live animals must travel longer distances, increasing pollution and congestion, as well as increasing overall product transportation costs (GÁMbaro et al., 2006). Many MS believe that these arguments are sufficient to justify spreading the costs of expensive veterinary meat inspectors among small abattoirs and meat processing plants to keep some of the value-added closers to the points of production rather than concentrated in cities closer to population centers. Training, education, and market knowledge are all essential. The government frequently subsidizes these occupations. Because corporations cannot benefit from talented workers who depart once the program is over, it is often assumed that the private sector would underfund this type of training. This market failure idea is especially visible in rural areas of the quality food chain, where small and micro companies predominant

CHAPTER 3

HOW SHELF LIFE OF FOOD IS RELATED TO FOOD SAFETY

CONTENTS

Shelf life is defined as the guide for a consumer of the period for which food is safe for consumption before it starts to deteriorate provided that any listed storage conditions have been met. The shelf life of a product usually begins from the time food is manufactured or prepared. Several factors influence the duration in which the products last before they spoil. Some of the factors include the type of ingredients, the process used in manufacturing, the type of packaging used, and how food is stored. The shelf life of a product is usually indicated in the labeling of the product. It is marked with a given date. There is significant evidence to prove that food shelf life is related to food safety (Goldman et al., 2002). The shelf life of a product describes the duration in which food retains its quality during storage. The pathogen content is used in describing the safety of food. Food safety is usually achieved by using a Hazard Analysis Critical Control Point system. Challenge testing and predictive modeling can be used to assess pathogen growth. However, it is important to note that there is an inextricably link between food safety and product shelf.

For food to be considered safe for consumption, it should remain microbial and chemically safe to eat, it should maintain its appearance, odor, flavor, texture, and taste. They should also meet the nutritional claims provided on the label. This places a great need for the shelf life of food to be accurate as it should ensure food safety. There are various things done to ensure that the shelf life of food is related to food safety. Some of the factors contribute to the safety of food though they are stored for a fairly long duration of time.

3.1 REGULATIONS RELATED TO SHELF LIFE

When storing food and determining its shelf life, there are regulations established by food boards of the given countries. For instance in Australia, the Australia New Zealand Food Standards Code is involved in establishing laws on regulations. The body has defined the composition and labeling requirements for food sold in New Zealand. The country has a code that defines a shelf life. It is listed in standard 1.2.5. The code requires that all packaged food with a shelf life of a few more than two years should be labeled with a date mark. This is linked to food safety as the code requires that food is safe up to and including the date marked. Among the options to be followed involves the adoption of the use of use-by date. The use-by date is mostly used for highly perishable food products that will present a safety risk if it is eaten after the given date (Galovičová et al., 2021). This means that the consumer should ensure that the products are consumed before the listed date. In most cases, consumers purchase products that have a long period before the -use-by date is attained. The state requires that food should not be sold past the use-by date. In some countries, if the company sells the products past the use-by date it is arrested. Consumers should also ensure that they do not eat food once the product passed the use-by date. Such foods should instead be thrown away as it is not safe for consumption. This means that foods that have surpassed their use-by date are a threat to human consumption.

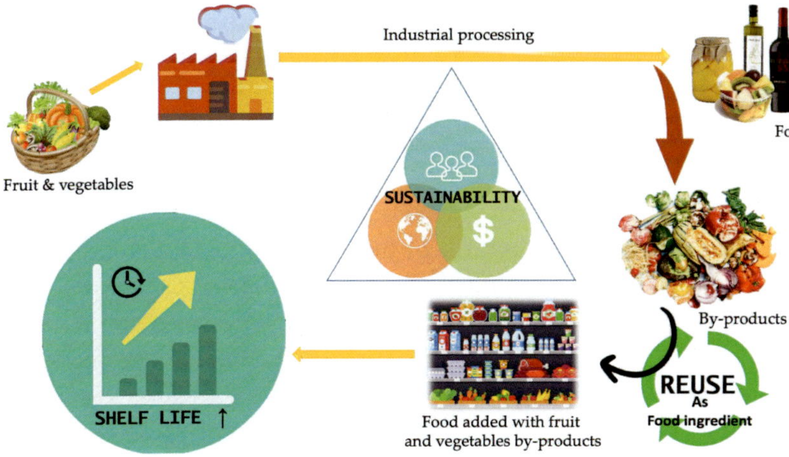

Figure 3.1: Regulation of food shelf life help to reduce the likelihood of food insecurity.

(Source: Nardella S, Conte A, et, al.,2022)

Another option involves companies using the best before. The best before date is usually used for food products that are semi-perishable. When the product reaches the indicated date, it has no safety risk but the quality of the product is. It is therefore not a threat to human consumption and not a threat to food safety. However, consumers should purchase food products that have not reached the best before date (Zhukov et al., 2018).

The other option is to use the baked on or baked for dates. Such dates are mostly used on pastries more so on bread products with a shelf life of fewer than 7 days. The baked form or baked-on date is usually placed no later than 12 hours after the bread was baked. The baked-on date is at times used by consumers to know how long the product will last before it starts to spoil. They can therefore fully utilize the product. This is also useful in minimizing the threat to food safety.

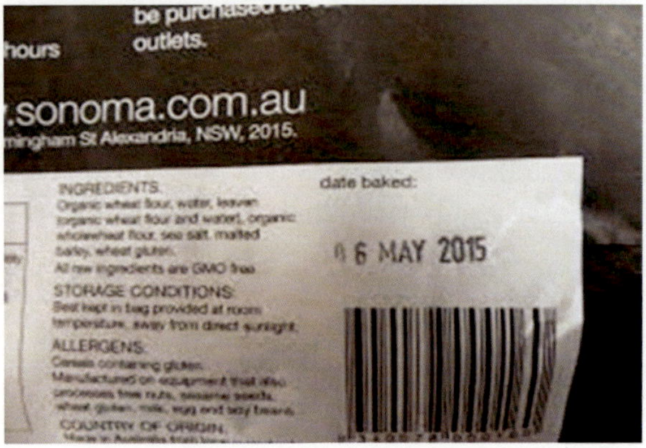

Figure 3.2: The "baked on" date is used for bread and other wheat products.

Depending on the manufacturer of the food products, there are different ways to write the date mark. In most cases, when the words "use by", "baked on", "baked for" and "best before" are used, they are followed by a date or reference to where on the package the date is located. Various requirements should be followed. It is a requirement that the date indicator should have at least the day or month for products with a shelf life of up to three months. For instance, the date indicator could state, "Best before 13 Jan." It could also include at least the month and the year for a product with a shelf life of more than three months (Weinbreck et al., 2010). A good example is JAN 30. The dates should also be expressed chronologically (day, month, and year) and numerically. However, the month can be expressed in terms of letters. They should generally not be encoded.

The requirement should also have a "packed on" date or the packer's code. However, they can only be used in addition to the date marks. There are different things listed in the code. They include the specification of storage instructions. This is usually listed on the labels. This will be read by consumers which will enable them to ensure proper storage of the food under the specified period that was indicated by the date marking. They can also ensure that all the storage conditions must be achieved in the distribution and retail system or at home. Those purchasing the product will also ensure the food is stored according to the storage instructions. The manufacturer is required to give instructions regarding the method of usage and storage of the product after the seal has been opened. This is very important for foods preserved under vacuum or modified atmospheres where the sealed packaging has a significant influence on the shelf life of the product. A good example is the usage and storage of sliced ham stored through vacuum packaging and has a month of refrigerated shelf life of one month. For such a product, once the package is opened, it should not be held for longer than 3-4 days.

3.2 CALCULATION OF FOOD SHELF LIFE

To ensure that foods are safely consumed within the required period, the shelf life of the product is calculated. Various individuals are responsible for the calculation of shelf life. For instance, an individual involved in the packaging and selling of food products that need date marks is legally responsible for calculating the period in which the product can reasonably be expected to keep without any changes in the quality of the product. The placed food label needs to have information on the shelf life and the storage instructions for the product to meet that shelf life. Calculation of food shelf life is usually the responsibility of the food manufacturer. There are cases where it is the responsibility of repackers, food retailers, supermarkets, and secondary processors (Villalobos-Delgado et al., 2019). They have formulated their method of calculating a shelf life. The formulation may take into consideration the try of ingredients used in making the product, the vulnerability of the product, and the expected shelf life when stored in certain conditions. Generally, it encompasses all the factors that influence the shelf life of the product.

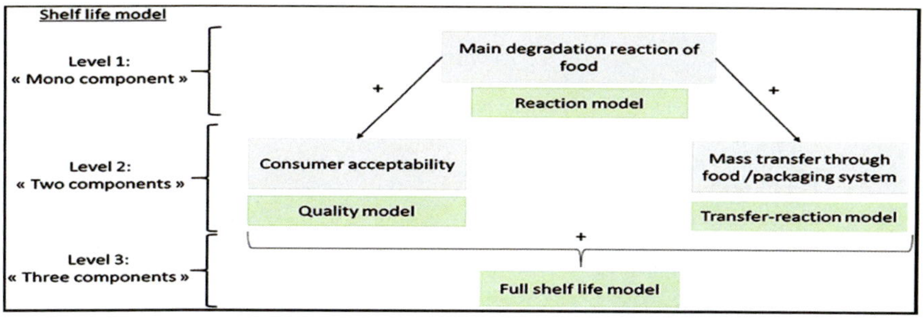

Figure 3.3: There are different formulas used in calculating food shelf life.

(Source: Coffigniez F, Matar C.,et, al.,2021).

3.3 FACTORS INFLUENCING THE SHELF LIFE OF FOOD PRODUCTS

Among the factors influencing the shelf life of a product are the people in the industry. Every person in the food production chain can influence food safety and quality. In most cases, the production may involve the consultation of a food control plan to boost their confidence. The food control plan is used in identifying and controlling hazards throughout the good chain. In the control plane, the role of each person is usually taken into consideration.

For instance, growers and primary producers are involved in the provision of raw materials. The quality of the products they provide can make production consistent generating a final product. Consistency of food quality can be maintained through the development of raw materials specifications. The role of suppliers includes packing materials or ice for chilling play a significant role in the safe and consistent supply of food materials. They also offer advice on different materials and how they affect the shelf life of the product (Wrona & Nerín, 2020). Manufacturers have the responsibility of determining the suitable shelf life of a product. The suitable shelf life of a product is determined based on a study that considers the stages of the product chain. Distributors are involved in the transportation of the food products from the manufacturer to the retailer. Distributors need to ensure that food is safely and securely stored in the warehouse. They should also ensure that the product is transported at the correct temperature. They should ensure that there are minimal or no delays while the packaging is not damaged. They should also take into consideration the possibility of food contamination from other carried goods being transported in the same vehicle.

Figure 3.4: Manufactures play a crucial role in determining the shelf life of a product.

The role of retailers is to ensure that they handle and store the product immediately before the product gets to the consumer. They are also required to store the product according to the guidelines stipulated by the manufacturer. The final link in the chain is the consumer. High emphasis is placed on the need for the consumer to be equipped with the correct information on storage conditions.

Concerning other factors influencing the shelf life of a product, there is no simple answer to the question. Naturally, foods tend to spoil with time. However, there is a considerable variation in the rates within which food spoils. Among the factors that influence the shelf life of a product is microbial growth whereby the growth of some bacteria, molds, and yeast in food may lead to food poisoning and food spoilage. The amount of time taken for microorganisms to affect food is dependent on their levels in food when introduced in food. Any further contamination of food during packaging, storage, and handling could also lead to microbial growth. Other factors that affect microbial growth that including the time and temperature of storage and the type of food. It is normal for moist foods to spoil at faster rates compared to dry foods. This is because moisture creates a conducive environment for the survival of microorganisms (Amani & Gadde, 2015).

The definition of the end of shelf life is made based on the number of microorganisms present. It is also made based on the recommended guidance. There are cases where the end of shelf life is determined by biochemical or sensory deterioration. Non-microbial spoilage is also a factor that affects the food shelf life. Under non-microbial spoilage, there are many ways in which the quality and the number of nutrients can be lost. However, this may not necessarily lead to the product being a threat to human health but it could mean that is no longer of the acceptable standard. When food products gain a significant amount of moisture, it can lead to loss of nutrients, rancidity, and browning. Dry foods become vulnerable to microbial spoilage when they absorb moisture. Any changes in the chemical state of food may result in off-flavor of food, loss of nutrients, browning, and changes in color. Some of the changes in food are brought about by light. Light-induced changes can lead to fading of natural colors, loss of vitamins, and ran acidity.

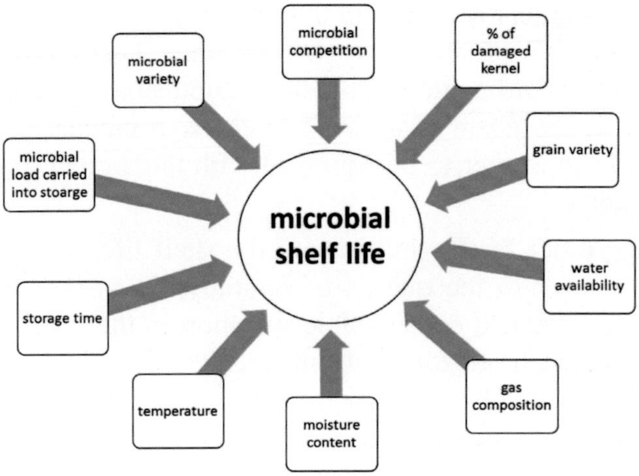

Figure 3.5: The shelf life of a product is affected by the type of microorganisms present.

(Source: Schmidt M, Zannini E, et al.,2018).

Any increase or decrease in temperature may speed up the rate or form of food spoilage. Any physical damage to food can result in spoilage. A good example is the bruising of vegetables and fruits. The bruise point creates an avenue for the entry of bacteria or other microorganisms. Any damages to food packaging may leave food vulnerable to both non-microbial and microbial spoilage. A good example is the presence of pin holes in tears or cans in plastic bags which allow microorganisms to enter food leading to the loss of moisture (Angourakis et al., 2015). Food spoilage can also be caused by insects and rodents. Food products can be tampered with by storing food near strong-smelling products. The strong smells will get into food tampering with their smell.

3.4 SHELF LIFE STUDY

When determining the shelf life of a product, a study of the shelf life is made to make accurate dates. A study on shelf life is an objective, methodological means to determine the period in which food products are expected to reasonably keep without any appreciable changes to their quality. Separate studies are done for each type of product. There are two main methods used in shelf-life studies. The first method is known as the direct method which is the most commonly used method. The direct method involves the product being stored under preselected conditions for some time longer than the expected shelf life. It also involves the product being checked at regular intervals to establish the time it begins to spoil. This leads to the development of a unique procedure for each product. The details of the steps to be followed and the types of decisions to be made are discussed.

In the indirect method, the approach makes use of accelerated storage and or predictive microbiological modeling to determine the shelf life of the product. There are different steps followed when determining the shelf life of the product using direct or indirect methods.

Steps involved in the calculation of shelf life through the direct method

Under the step-by-step determination of shelf life, several decisions should be made during the study. When making such decisions, there may be a great need for one to seek expert advice. The first step is to identify the main cause of food spoiling or becoming unsafe. It is important to note that each product has its own set of factors that may limit its shelf life of the product. There is a list used in the identification process (Winterhalder et al., 2015). The list is often used as a starting point in helping identify the possible ways that the product will lose its quality or safety. At the same time, the list is used in identifying factors that can be vital in prolonging the shelf life of the products.

During the decisions making process, it is very important to consider the entire process. The process starts from the purchase of ingredients and packaging materials right through to the final step, which is the end-use by the consumer. The time of the year is also taken into consideration as some products are known to deteriorate faster in summer than winter ad a result of higher temperatures. The shelf life of a product is also determined by the related spoilage.

Figure 3.6: The quality of raw materials will affect their shelf life.

They include raw materials whereby the quality, consistency, level of contamination, and storage of raw materials may affect the final quality of the product. The product makeup also affects food spoilage where the ingredients use and how they behave when combined will influence the type of and how many spoilage organisms will grow. If there are any changes to the proportions of ingredients or the type of ingredients will affect their shelf life. The water activity of a product is defined as the amount of water in a food product that is available for use by microorganisms (Wibowo et al., 2018). For this reason, water creates a conducive environment for the survival of microorganisms. The addition of sugar, salt, and other ingredients will reduce water activity. By nature, jam is moist food but the large amounts of sugar it contains means only a small amount of water will be used by microorganisms.

The pH of the food is defined as the measure of alkalinity or acidity of the food. The pH of the food is known to influence the type of microorganisms that will thrive in a given food and also grow. A good example is fresh mussels are known to have a very short shelf life while mussels in vinegar marinades have a relatively long shelf life as the acidic environment limits microbial growth. The availability of oxygen in food is achieved by removing air from around or by the use of vacuum packaging or through atmosphere packaging that helps extend the shelf life of the product. It is important to note that some microorganisms can grow in environments without oxygen. For this reason, the production process needs to be controlled to control the microorganisms. During the processing of the product, food safety can be ensured by maintaining high standards of hygiene. This is because the methods of packaging limit the growth of spoilage organisms that require oxygen. They do not kill microorganisms but limit the growth of organisms that grow in the absence of oxygen. Examples of such kinds of bacteria include bacteria causing slime, listeriosis, or botulism.

The use of chemical preservatives correctly is useful in controlling the growth of microorganisms. It is important to note that they have more than one role. A good example is sulfites that slow spoilage and also prevent the browning of dried fruits. The amounts and types of preservatives that can be used in food are listed in the Food Standard Code.

3.5 PROCESS RELATED SPOILAGE

The food safety of a product and its shelf life can be affected by the processes used during production. There are processes that smoking, mixing, salting, heating, chilling, fermenting, and cooling. They can also include dehydration, freezing, and heat sterilization. The type of processes used in food production can alter the shelf life of the final product. A good example is UHT milk which is a sterilized product and is known for its very long shelf life compared to pasteurized milk (Uebele et al., 2013). This is because the heat treatment is greater. When creating several products, consistency is very important. It ensures that all products have the same shelf life, flavor, and appearance among others. It is also important that the quantities, quality, and processing ingredients are the same. In most cases, companies have a written flow chart or recipe that gives the exact details of the procedure to be done at each step more so the times and temperature.

Any minor variations in the time and temperature can lead to failure to target spoilage and food poisoning organisms. Packaging is also crucial. When selecting the type of packaging, it is important to note that the type of packaging used is useful in protecting the product from contamination during the subsequent steps such as distribution, sale, and domestic storage. Poor packaging can lead to the contamination of various products. If they are not well-produced or stored hygienically (Calligaris et al., 2016).

Storage of products can also affect their spoilage where the temperature of storage can slow down the growth of microorganisms which are important determinants of food safety and quality. Other factors that should be taken into consideration include physical handling, light, humidity, placement near other products that could taint the food, and protection from insects, birds, and rodents. Temperature data loggers are used to ensure correct controls are maintained as fluctuations in temperatures affect the shelf life of perishable products. Once all the factors that could lead to food spoilage have been identified, all the necessary measures should be put in place to ensure that food safety is maintained.

There is a great need for food production companies to select suitable tests when determining the quality and safety of the product. It is important to note that one test may not be appropriate for all products. For example, there may be a method chosen in testing raw meats for the number of lactic acid bacteria but it may not be used in testing fermented raw meats for the organisms. Among the tests done on food include checking for food poisoning organisms such as listeria to verify product safety. When the product is to be subjected to laboratory tests then the manufacturer needs to check the laboratory to ensure that it is accredited for those with IANZ. There are cases where storage trials are conducted by a laboratory then there is a need to see records of storage temperature stability.

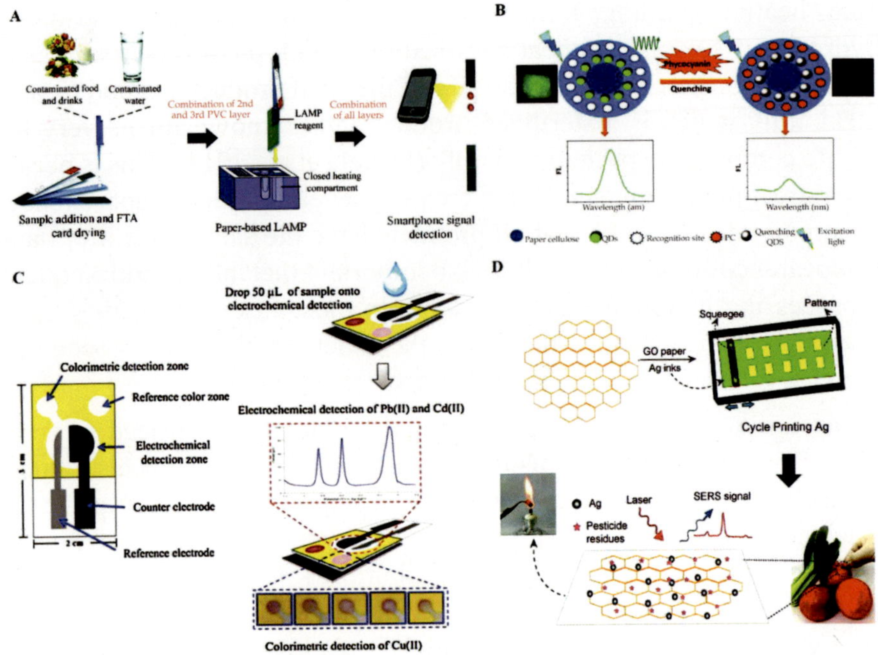

Figure 3.7: There are different kinds of tests used in determining food safety.

(Source: Choi JR, Yong KW. et, al., 2019)

Generally, tests are usually divided into four categories. The first category comprises sensory evaluation. Under sensory evaluation, it involves the assessment of food's texture, flavor, appearance, and smell. This kind of evaluation is used in monitoring and recording obvious changes that occur over time. For this reason, it is used in determining the shelf life of food. It usually involves food being assessed under the conditions at which it is designed to be stored and consumed. In an ideal case, it is done by a trained panel that makes use of a recognized evaluation method (Turner et al., 2020).

Microbiological tests are also done on food. Such kinds of tests are usually done when evaluating food safety and food quality. The tests may be done when estimating the changes in the number and type of spoilage organisms such as yeast, bacteria, and molds and occur throughout some while. Certain standard plate count levels apply to various ready-to-eat food groups. There are methods used in determining the standard plate count of certain foods. To ensure food safety, it is important to ensure that food poisoning organisms are not found in food samples. The type of tests done is dependent on the type of product. Some microbiological guidelines and standards give guidance on the type of organisms and their number that can be said to be acceptable or unsafe in food.

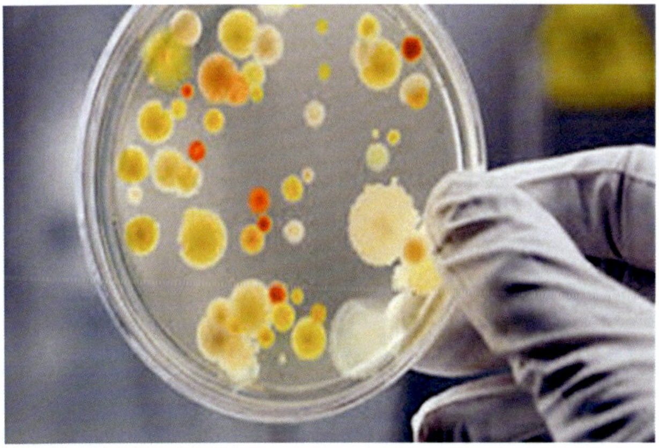

Figure 3.8: Microbial tests are useful in evaluating food quality and safety.

In New Zealand, there are criteria followed when testing food. The microbiological guidelines criteria outline the guidelines for the microbiological examination of ready-to-eat foods. The criteria listed are useful in providing guidelines to the food industry when setting acceptable levels of microorganisms in food. In the NZFSA it is recommended that the food industry set its own more stringent microbiological standards. In some European countries, there are standards set concerning the microbiological criteria for *L.monocytogenes* at the end of shelf life (Castrica et al., 2021). When determining the criteria, there are guidelines useful in providing the needed assistance used when determining the shelf life of food that can support the growth of *L. monocytogenes*.

There are chemical tests done on food. These tests are useful in detecting any changes in the quality of the product throughout its shelf life. Some examples of instrumental chemical tests include total volatile nitrogen, headspace gas analysis, and pH. Food can also be subjected to physical tests. Such kinds of tests involve the measuring of product texture, examining the package, travel tests, and also determining the worst, best, and average retail conditions. A travel test is used in identifying any hazards involved during transportation and handling of food. It usually involves the transporting of food products through the expected distribution and storage chain. The food products are examined at various points and at the chain which is a very important step. Records of temperature at present times are entered into a data logger. The records are used for later analysis. The main goal is to ensure that the experimental design should mimic real-life practices such as the predicted transport temperature to retail outlets, during commercial control, consumer storage, and consumer purchase and transport. The company usually selects the type of test that will suit its products.

Step 3 involves planning the shelf life study. Certain points are taken into consideration when developing a detailed shelf life study plan. They include the type of test to be carried out and the duration in which the study will run and the frequency with which the test will be carried out. In the plan, one should include the actual sampling dates. Collection of the samples is should be done at the beginning, at the target endpoint, and on about three occasions in between. Sampling can also be done beyond the target to confirm the endpoint selection. With regards to the number of samples that will be tested at each time, it is required that at least triplicate packs of products should be tested at each sampling. The manufacturer has to establish the number of samples needed for the whole study period and the time in which the study will run. In an ideal situation, the test is done during the season most likely to cause problems. In most cases, summer is the season when these tests are done (Timakova et al., 2020).

Manufacturers should conduct the study more than once. This is useful in taking into account the variability of the product. It is also required that the product, process, and packaging be the same as the intended use for the final product. They should also ensure that they keep a written record of everything done. The written records are useful in interpreting results. This makes the study fully planned and timetabled.

Step 4 involves the company running the shelf life study. During these studies, the samples are stored under the same conditions as the normal production samples from manufacturing to consumption. In cases where this is not possible, the samples should be stored in known humidity and temperature. In such a situation, they should be regularly checked and recorded. Time set out in step 2, is used to direct the selection and testing of samples as per the test decided in step 1. Results of the tests should be recorded.

Step 5 involves the determination of shelf life. During the study, there is a point reached where the product no longer meets the quality standard. All the information that has been recorded and observed is used in deciding how long the product can be kept and still have acceptable quality and safety. It is important to note that maximum storage times for safety and quality may not be the same. The most suitable shelf life of a product is the shortest. If the result of the conducted tests does not make any sense, the test should be repeated. If the test is repeated but they still don't make sense or the results are variable then the used ingredients should be checked with major emphasis being placed on the analysis of quality and the processing are similar for all batches (Sutton et al., 2016). This allows an analysis of what is causing the variability which can, later on, be fixed and sampling repeated. This enables an estimation of food shelf life based on the ideal storage conditions. This is an allowance for the real world where the storage conditions are variable and there may an occurrence of product abuse. Manufacturers should ensure that the shelf life of the given product should be reasonable and not ideal. They should also have a safety margin. This can limit the possibility of product abuse to some extent by specifying the much-needed storage conditions for the product while limiting its distribution.

Monitoring of the shelf- life of food samples being tested for factors that have been highlighted as the most important for the given product. These factors include acidity, level of spoilage organisms, and loss of flavor. The used samples can also be taken from various points within the retail and distribution system. When the results of the test show that the preliminary shelf life is inappropriate, they should be adjusted. Studies on the shelf-life of the product should be repeated if there are any changes made in the processing and production environment. The product can develop a longer shelf life through the identification of limiting factors in the shelf life study, modifying the, and finally repeating the study. Any complaints made by consumers concerning product failure should be investigated if the complaint is made before the expiry date. This is useful in identifying a recurring problem. In other cases, it may indicate the need to recalculate the shelf life of the product.

All the records made when designing and carrying out the shelf life study are useful in the evaluation of customer complaints, troubleshooting, distribution, and production problems. They help review the shelf life of the product. Manufacturers need to ensure that all the results of tests are written down and kept in a safe and accessible place. The product is continuously monitored to ensure it is safe and of good quality throughout the product's entire shelf life.

3.6 INDIRECT METHODS

Other than the direct methods, indirect methods have been used to attempt to predict the shelf life of a product without running a full-length storage trial. For this reason, they are used to dealing with food products that have a long shelf life. There are two classifications under indirect methods. They include accelerated shelf-life studies. Such studies are composed of a trial period that has been shortened by the deliberate increase in the rate of deterioration. The rate is increased by increasing the storage temperature. The results from the study are used in estimating the shelf life under normal storage conditions. The other classification is predictive modeling. The predictive models are mathematical equations that make use of information from databases to predict bacterial growth in foods under defined conditions (Sewald & DeVries, 2003). These models are mostly used in calculating the shelf life of a food product. The studies have a major emphasis on the information on the changes that occur in the product when it deteriorates. For this reason, the properties of the product and also its packaging is involved in the calculations. In most cases, the predictive models are specific to particular types of organisms.

The United States Department of Agriculture Pathogen Modeling Program is a good example of a protective modeling system. Other models other the same department include the Growth Predictor and Forecast which is a paid service by Campden and Chorleywood Food Research Association. These models are useful when used as a first step in the evaluation of a product's shelf life. However, it is very important to note that information from ing modeling programs needs to be verified by shelf-life trial or challenge testing (Chang et al., 2022). There are consultants in food safety and technology who assist with specific predictive modeling trials or problems. There are various consultants in the food industry.

3.7 CHALLENGE TESTING

Challenge testing is very important in ensuring food safety. Challenge testing is often used in assessing whether the product formulation and storage conditions of food can control the growth of pathogens if they are present in the designated shelf life. The procedure is followed during the inoculation of the product with relevant microorganisms and incubation of the product under controlled environmental conditions to assess the risk of food poisoning. It can also be used when establishing stability. For this reason, there are particular scenarios where challenge testing is used. The preservation effect in food is brought about by a combination of various factors none of which is sufficient to control the safety of food obits own. Challenge testing is often used when the effect of such a food control system on specific pathogenic microorganisms cannot be predicted from the literature. In this case, challenge testing is done to evaluate the safety of the product. Under the standards of food production, there are guidelines for various tests.

3.8 GUIDELINES FOR MICROBIOLOGICAL EXAMINATION OF READY-TO-EAT FOODS

Under these guidelines among the areas of great importance are the standard plate count levels applicable for ready-to-eat foods. Depending on the groups of food, there are different standard plate count levels. For instance for meat brawn is 2, smoked meats (chicken, hot pork, pastrami) 2, sliced salami and other fermentation meats 3, sliced meat (luncheon, corn silverside, ham) 2, sausages, saveloys and frankfurters at 1, minced patties that are cooked at 1, kebabs without salad at 1 and cold meat and poultry at 2.

Foods classified at level 1 are those considered RTE foods in which all components of the food have been cooked in the manufacturing process or preparation of the final food product and microbial counts should be relatively low. The term level 2 is given to RTE foods that contain components that have been cooked and then further handled during storing, slicing, and mixing before preparation of the final food or where minimal or no cooking processes have been used. The classification of level 3 does not apply to SPCs. It applies to foods such as vegetables and fresh fruits such as salad vegetables, fermented foods, and foods incorporating them. Such foods are expected to have an inherent high plate count because of the normal microbial flora present (Sumner & Jenson, 2011). It is important to note that if a specific RTE food is not categorized in a given standard, the individual should use their professional judgment to make a decision where the product would fit based on the type of product and processing that has been received.

3.9 DETERMINATION OF SHELF LIFE OF FOODS CAPABLE OF SUPPORTING *LISTERIA MONOCYTOGENES*

When determining the shelf life of food products, it is important to take into consideration whether the food product is capable of supporting the growth of *Listeria monocytogenes*. Among the most dangerous pathogen concerning chilled processed food is *L. monocytogenes* as it is capable of growing at refrigeration temperatures. The occurrence of listeriosis is a threat to food safety as it causes various illnesses. As a pathogen, it has serious health consequences for susceptible groups of people more so those who are very young, elderly, pregnant, and those who are immunocompromised. In most cases, healthy people will suffer from a non-invasive gastrointestinal illness. However, they can be affected by the more serious outcome. Determination of when the growth of such pathogens occurs is not very simple as several factors affect it. These factors include storage time, competing for flora, acidulant, pH, and temperature. There are various characteristics of *L. monocytogenes*.

3.9.1 Characteristics of *L. monocytogenes*

Among the characteristics of *L. monocytogenes* is that they are capable of both aerobic and anaerobic growth such as in a vacuum pack. They have a minimum pH of 4.3 to facilitate growth, a minimum temperature of -1.5 degrees to facilitate growth a decimal reduction time of D60 degrees Celsius, 5-10 min D70 degrees approximately for 10 seconds.

Foods made of mixed cultures have noted poor competition. Refrigerated foods as in the case of vacuum-packed meats with low levels of competing organisms can provide favorable conditions allowing the multiplication of Listeria. Special consideration is taken when determining the shelf life of high-risk foods. Some of the factors include their ability to support the growth of *L. monocytogenes*, whether they are sold ad ready-to-eat, storage involves refrigeration, and are stored for long periods. There are various issues used to help manufacturers determine whether products are of the greatest concern. They include the availability of *L. monocytogenes* being present in raw materials, the ability of an organism to be destroyed in any of the processing stages, the likelihood of the product being exposed to post-process contamination, the likelihood of the finished product allowing the growth of *L. monocytogenes* if the product has a long shelf life and whether the product will be subjected to a process by the consumer that will lead to the destruction of *L. monocytogenes*.

Some food products said to bear high risks of *L. monocytogenes* include high fat, unpasteurized milk, smoked seafood, meat spreads, paté, frankfurters, deli meats, other dairy products, and soft unripened cheeses. Many surveys have established the times required for *L. monocytogenes* to grow through one log in some ready-to-eat foods in the temperature range of 4-12 degrees Celsius (Dawson et al., 2018). They include cold-smoked seafood which is 43.2 h, vacuum-packed deli meats at 48 h, and ready-to-eat crustacea at 48h. When determining the shelf life of high-risk products, the manufacturer considers whether the product can be put in place of any controls. The controls include post-packaging treatment of food products to destroy Listeria, preventing the contamination of foods that allow the growth of Listeria, the establishment of acceptable storage times to produce extensive growth, and prevention of the growth of Listeria in foods through the addition of antimicrobial agents or the use of modified atmosphere packaging, the provision of advice to consumers on how to thoroughly cook food and also providing advice to susceptible groups like labeling food as unsuitable for consumption by the susceptible groups.

3.9.2 Growth of *Listeria Monocytogenes* in Food

When storing food for long durations of time, the likelihood of the growth of *Listeria monocytogenes* affects food safety. Various data have been collected concerning different types of cheese, vegetables, and meat products but little knowledge of other things. The available data has been extracted from a project conducted by ESR for the Ministry of Health to present health protection officers with useful information on the various kinds of food that will support the growth of *Listeria monocytogenes*. There are various kinds of information concerning data. The worst case is when two isolates have been tested and one of them grew while the other didn't. This means that the data of the product that grew is the only one included. This is an indication of adaptation. For instance, when the product was previously exposed to low pH, it can result in a subsequently enhanced survival if the product is under acidic conditions (De Corato, 2020). The results, however, are an indication of a change in the rate of death or growth of the product and therefore a matter of degree.

This means that if *L. monocytogenes* are reported to die in food then the adapted strains will still die at a slow rate. The term growth in this case is used to refer to an increase in number while survival means that there was neither a significant decrease nor an increase in numbers. The term decline means that there was a decrease in the number of organisms present under the conditions in the food. There are cases where some of the products will appear in more than one place. A good example is when *L. monocytogenes* grows on food at 10 degrees but not at 3 degrees Celsius. It is important to note that most of the listed temperatures are those at which the experiments were conducted usually at refrigeration or mild temperature abuse values. It is generally safe to say that food was found to support growth at 3 and 6 degrees Celsius and will do the same at 4 and 5 and temperatures higher than 6°C.

3.10 SHELF LIFE AND INITIAL NUMBERS

In some of the experiments done, they review the measured growth of Listeria already existing in purchased foods hence natural contamination. In most cases, the initial counts are very low and the subsequent growth is slower at refrigeration temperatures than expected from experiments where cells were added to foods. This is known as the inoculum effect which is quite well reported for listeria. There are two main phases of bacterial growth namely the lag phase and the exponential phase all exhibited attaining the maximum number.

During the lag phase, bacteria take their time to adapt to the new environment. The lag may be very long when placed in sufficiently low storage temperature that will vary with food. The lag phase will be so long that even if the organism was theoretically able to grow in food, no growth will be exhibited at a relatively short shelf life. Studies show that in products such as minced beef characterized by short shelf life during refrigeration, no growth of Listeria is likely to occur before the product becomes grossly spoiled, no growth of Listeria is likely to occur before the product becomes grossly spoiled and is therefore unlikely to be sold or eaten as it poses a threat to food safety (Derkach et al., 2020). During experiments, when the storage period was shorter than the lag phase *L. monocytogenes* was reported as not able to grow. However, other experiments with extended storage periods with the same goods will record as allowing growth. This is very essential in explaining why similar food items may appear in more than one place in certain categories.

3.10.1 Extended Shelf Life Foods and *Clostridium Botulinum*

By nature, *Clostridium botulinum* is known to produce heat-resistant spores with the ability to survive in pasteurization. They can also survive at cooking temperatures. Such organisms are known to produce potent toxins in food. Food is potentially hazardous to consumers and a threat to food safety when there is a potential for spores of *C. botulinum* to present in food and the conditions will allow the generation and growth of spores. In most cases, the outbreak of botulism poisoning has been noted in canned and bottled vegetable products or meats. In such a case the spores were present in raw food. Heating is known to kill competitive organisms. However, there are spores known to survive and grow during storage. It is expected for spores of *C. botulinum* to be found in raw foods harvested from water and land. Such organisms can be eaten without necessarily causing any harm. This occurs when conditions can facilitate the growth and toxin production in food that botulism poisoning will occur. This leaves large populations of people at risk of botulism. The symptoms of the disease may vary from mild disease to an illness that can prove fatal within 24 hours. By nature, botulism is known to have a high fatality rate and of the cases reported about 80% are hospitalized and need intensive support therapy.

Some examples of foods involved in botulism outbreaks include mascarpone cheese, meats (pies, pates, fermented, canned, cured), fish and seafood (canned, smoked, vacuum packed, raw), bottled fruits (peaches, apricots, pears), vegetables (sautéed onions, garlic oil, baked potatoes), tomatoes and tomato juice, olives, roasted eggplant in oil, canned and bottled vegetables such as marinated mushrooms and peppers.

There are two main types of *C. botulinum* that affect human beings. The first group is composed of Proteolytic strains that produce neurotoxins A, B, and F. They are capable of anaerobic growth such as in vacuum packing. They have a minimum pH of 4.6 that facilitate growth, a minimum aw for 0.94 growth, a minimum growth temperature of 10 degrees Celsius, decimal reduction times at 121 degrees Celsius at 0.2 min, and recommended commercial heat treatment of 3 min at 121 degrees Celsius. Group 2 is made up of Non-proteolytic that produce neurotoxins B, E, and F (Sarkar et al., 2022). They are capable of facilitating anaerobic growth in a vacuum pack, they require a minimum pH of 4.7 for growth, a minimum aw of 0.97 for growth, the minimum growth temperature of 3.3°C, decimal reduction times at 90° C at 1.1 min, and a recommended commercial heat treatment of 10 min at 90° C.

3.11 CHILLED FOODS

When ensuring food safety a major concern has been concerning chilled vacuum-packed foods. This is because there are types of *C. botulinum* that can grow at refrigeration temperatures that can go as low as 3.3 degrees Celsius. Certain foods have specific control factors that can prevent growth. They include low pH of high salt. This means that they do not present any risk. In other food, a combination of preservative effects is used in controlling the growth of *C. botulinum*. There are standards set by food committees to ensure food safety. A good example is the UK Advisory Committee on Microbiological Safety of Food. The body recommends chilled foods with a shelf life of a few than 19 days in addition to chill temperatures of less than 8 degrees Celsius. There are certain controls used singly or in combination. They include heating to 90 degrees Celsius for 10 min, minimum salt level of 3.5%, Aw of less than 0.97, and a combination of preservative and heat factors that can be shown to prevent growth and toxin production of *C. botulinum*.

For foods stored at ambient temperatures, many factors can contribute to the safety of long shelf-life products. One can incorporate a heat process to eliminate the heat hazard at high temperatures to achieve botulinum heating. A good example is 121 degrees Celsius for at least three minutes for bottled or canned foods. One can also use a combination of processes when controlling the hazard raw material control. This is achieved through the selection of the best quality raw materials to ensure levels of *C. botulinum* are not excessive (Roccato et al., 2017). They can also be heated to 90 degrees Celsius for 10 min, a product can be formulated to prevent the growth of *C. botulinum*, the addition of acid or fermentation, drying or salting, and the addition of preservatives.

Consumers are often advised to pick products carefully. This is because some products rely on effective handling. The consumer plays a big role in ensuring that they remain safe until they are consumed. Consumers get much information from the labels placed on the products. Information could include "once opened keep refrigerated" or "consume within". In as much as food-associated botulism is rare, with the growth of novel food products there is an increasing market for vacuum and modified atmosphere packaged chilled foods. They are also affected by the availability of food ingredients from any part of the world. There is also a great need for the manufacturer to consider *C. botulinum* in their Hazards assessment when establishing the shelf life of food products. Other factors should also be taken into consideration depending on their significance.

3.12 STORAGE

A great part of the relation between food safety and food shelf life is storage. This is because how food is stored will influence its ability to last for the outlined shelf life. Various things should be taken into consideration when looking into the storage of food products. Storage is the role played by consumers and also manufacturers. Manufacturers have to ensure that food is properly stored before it reaches the consumer. They both need to beware of the temperature danger zone. The temperature danger zone is the zone in which food poisoning bacteria can grow at a fast rate. The zone is characterized by temperatures between 5 and 60 degrees. High-risk food should be kept out of such temperature zones.

Handling of high-risk foods should be done with high levels of care. This is because food poisoning bacteria can grow and multiply faster in high-risk foods compared to other types of food. Some high-risk foods include raw and cooked meat such as minced meat and chicken. It also comprises food that contains raw and cooked meat such as lasagne, curries, and casseroles. Dairy products such as dairy-based desserts and custard-like custard tarts and cheese are high risk. Other foods include eggs and egg products such as mousse, small goods such as salami and ham, seafood such as fish stock, fish balls, patties, seafood salad and stew containing seafood, pasta and cooked rice, prepared salads such as rice salad, pasta salad, coleslaw, prepared fruits salads and ready-to-eat foods such as pizza, roll, and sandwiches containing mentioned foods (Pratiwi et al., 2015).

Other groups of foods that can be classified as high risk are those that come in packages. For such kinds of food, once they have been opened they should be properly handled and stored correctly. For a significant portion of food, their shelf life is sustained by placing them in the fridges which will maintain it at a low temperature. When storing foods in the fridge, the fridge temperature should be at 5 degrees Celsius while that of the freezer should be below -15 degrees. A thermometer is used in checking the temperature of the fridge. There are food items whose shelf life is sustained by storing them in the freezer. For such foods, when shopping one should purchase chilled and frozen foods at the end of the trip. They should also take them home as soon as possible to continue with storage. If it is a hot day or the trip will take longer than 39 minutes then one should take an insulated cooler bag or ice pack to keep the frozen foods cold. It is very essential to ensure that hot foods and cold foods are kept separate. One is at home, they should put the frozen and chilled foods into the fridge immediately. It is important to ensure that foods stored in the freezer are frozen hard.

Under proper storage of cooked food, once food has been cooked it should be allowed to cool. Hot foods should be placed in shallow dishes or separated into smaller portions to help cool food as quickly as possible. Very hot foods should not be placed in the refrigerator. Instead, it should be allowed to cool such that steam is no longer rising from the food before it is placed in the fridge. Another major part of food safety is ensuring that thawed food is not frozen. Food poisoning bacteria can thrive in frozen food while it is thawing. For this reason, one should avoid thawing frozen food in the temperature danger zone (Peleg et al., 2017). Defrosted food should be placed in the fridge until it is ready to be prepared. When a microwave oven is used in defrosting food, the defrosted food should be cooked immediately. With regards to handling food, one major rule is that one should avoid refreezing thawed food. There is a higher likelihood of food that has been frozen a second time save having high levels of food poisoning bacteria. The type of risk incurred is dependent on the condition of food when frozen and how food is handled between thawing and refreezing. For raw foods, once they have been thawed they should never be frozen.

Handling of food also requires raw food to be stored separately from cooked food. They should be stored separately in the freezer. When stored together, there is a possibility that bacteria from raw food can contaminate cold cooked food and the bacteria can multiply to very dangerous levels if it is not thoroughly cooked raw food should also be stored in a sealed or covered container at the bottom of the fridge. Raw food should be kept below cooked foods to avoid liquids such as meat or vegetable juices dripping down and contaminating the cooked food. Storage containers also affect food safety. One should choose strong, non-toxic storage containers. One should also ensure that they are clean and in good condition. Such containers should only be used when storing food. They should be covered with tight-fitting lids, plastic film, or foil to minimize risks of contamination. Once food has exceeded its shelf life, it should be thrown away. The same case applies to food that has spoiled even before the expiry date has reached.

CHAPTER 4

REGULATIONS RELATING TO SHELF LIFE

CONTENTS

Shelf life is the length of time that a commodity may be stored without becoming unfit for use, consumption, or sale. In other words, it might refer to whether a commodity should no longer be on a pantry shelf (unfit for use), or just no longer on a supermarket shelf (unfit for sale, but not yet unfit for use). It applies to cosmetics, foods and beverages, medical devices, medicines, explosives, pharmaceutical drugs, chemicals, tires, batteries, and many other perishable items (Omotilewa et al., 2018). In some regions, an advisory best before, mandatory use by or freshness date is required on packaged perishable foods.

Regulations on shelf life relate to the safe production of food and the protection of consumers. To ensure maximum freshness, companies and industrial infrastructures apply a wide range of packaging materials and unique processes to package our foods. This is aimed at producing packaging products that protect product quality, extend shelf life and help maintain the optimum quality right up until the last bite. The safety of products regarding shelf life is not only applied in food products but also other commodities e.g. in the beauty and body fair industry. The shelf life of cosmetics varies from makeup item to makeup item, depending on the type of product (i.e. eye shadow, lipstick, perfume). It is not difficult to research the expiration date of a product or compare the expiration date with that of another similar item in one's collection. A little research goes a long way toward eliminating uncertainty and stress-free shopping for food products and cosmetics for consumers.

Important to note, Shelf life is not a method of food preservation. Shelf life is instead the length of time that packaged foods, beverages, drugs, and other products will retain their quality, safety (in terms of microbiology), or nutrition when properly stored under normal conditions. The definition of shelf life includes the specific storage requirements for a product to maintain quality. The term is usually used as a partial synonym for food safety. Manufacturers who do not meet the standards required can be charged by local authorities, or lose customers by providing a substandard product. Goods that are stored in containers and packages (pouches, boxes, crates, and bottles) are accompanied by a date of production, batch number, and expiry date.

Figure 4.1: The expiration date is closely monitored by agencies to ensure consumers are sold safe and non-expired goods.

Shelf life is legally defined in a very narrow context as the time that an item in a shop may be displayed for sale, or the time it (any perishable or potentially perishable item) may be stored, during which the defined quality of a specified proportion of the items remains acceptable under expected (or specified) conditions of distribution, storage and display. Perishable food items have a minimum allowable shelf life. This regulation is a consumer protection measure to ensure that perishable products are safe when they are consumed.

It is essential to ensure food is stored and handled correctly. The shelf life of a product can be divided between the Period of Use (PU) and the Period of Validity (PV). The PU refers to the length of time during which a food item can be used while still being fit for consumption. In other words, it is the maximum length of time from when a food item is fit for consumption until it deteriorates significantly. The PV refers to the period for which a proper quality level is assured in terms of sensory, nutritional, and microbiological characteristics. Then there is the factor of expiration dates. Expiration dates on consumer products aim to guarantee consumer protection.

The concept of the expiration date is related but legally distinct in some jurisdictions. If a sell-by date is provided, the product should not be sold or consumed after the given date. If an open dating system is used on food products, discard the product if it is not consumed within 21 days after opening the container. Shelf life is the recommended maximum time for which products or fresh (harvested) produce can be stored, during which the defined quality of a specified proportion of the goods remains acceptable under expected (or specified) conditions of distribution, storage, and display. If the cans look okay, they are safe to use. Discard cans that are dented, rusted, or swollen. High-acid canned food such as tomatoes, grapefruit, and pineapple can be stored on the shelf for up to 18 months.

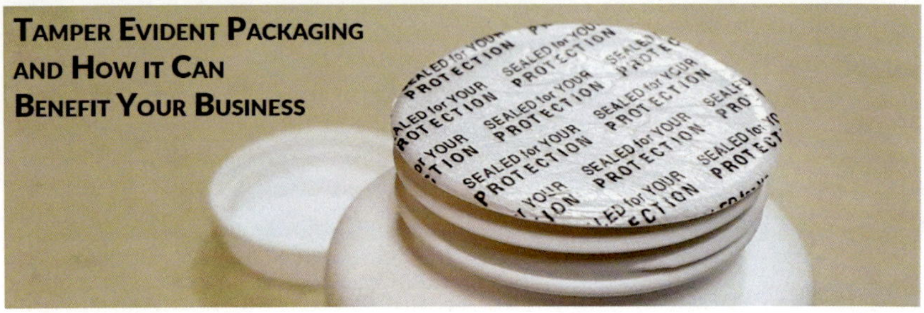

Figure 4.2: Goods should always be protected from tampering to mitigate the entrance of germs.

For instance, a can that is swollen, dented or bulging, or leaks when its contents are poured out, should be thrown out. That's because the quality of the food can deteriorate and make you sick. If a can has a dent or crease, press it in to see if it is soft. If the dent stays in place when you press it, then throw the can away because air may have entered the can and led to bacterial growth. One can also check the canned food for rust or corrosion by running fingers over the tin. If there are areas that are rough, rusted, or corroded, toss them immediately because they can easily contaminate what's inside the can. This is the essence of understanding shelf life.

Product shelf life varies according to the manufacturer's instructions. Depending on the products, the applicable laws regulating shelf life differs. This is based on are based on laboratory studies known as microbiological challenge testing. Food labels are made according to FDA standards. The standard shelf life is largely based on product test results taken at the time of manufacture, regardless of how long it has been since manufacture. For example, foods with a natural short shelf life might have a shorter than expected label lifespan after baking, when frozen and thawed, or when left out on the counter too long (O'Brien & Bentley, 2015). The labeling requirements ensure that consumers can purchase foods with the solid expectation that the food will not cause illness for an increasingly long period.

Any material that has been opened, used, or tampered with will have a shorter usable shelf life than an unopened container that has been stored under ideal conditions. In general, the manufacturer will provide an expiration date for the agent indicating the date until which time, if properly stored, it is deemed safe for use. In many cases, however, an opened agent can be useful beyond its expiration date and may still be used for training purposes but should be destroyed after that. To determine the effectiveness of an agent past its printed expiration date, users should consult with their local civil support team (CST) representative or contact the product manufacturer directly.

4.1 GENERAL PRINCIPLES

The general principles on regulations relating to shelf life include the shelf lives of foods and other products are the lengths of time for which they are safe to consume or use, under specified conditions. Requirements for shelf life are often specified in law or regulation. Generally, procurement organizations will not buy food with less than 75 percent of its shelf life remaining. There is a close relationship between shelf life and product date, but they are not the same thing. For example, bread that has been frozen may have a brand new package date but will have expired in terms of edible quality many months ago! (Nicoli, 2012).

To ensure safety, quality, and public confidence, it is of prime importance that foodstuffs are properly handled, preserved, stored, and transported. The general principles for the regulation of shelf life of foodstuffs can be summarized as follows:

- Shelf -life is determined in all cases by the potential hazard to humans and animals.
- The shelf-life shall be based on relevant data about the stability of the product, including packaging, both at normal and appropriate abnormal conditions under which the product may be stored.
- The shelf-life shall take into account any post-treatment processes that could affect its compliance with this regulation.
- concerning primary and secondary packaging, labeling indicating ‘best before...’ or similar expressions, may be used if these are justified by a reasonable assurance of satisfactory quality for the expected shelf-life, provided that immediate and effective recourse is possible in case of variations.

- Relevant legislation concerning traceability should also apply
- Microbiological and physico-chemical parameters
- Hygienic production and preservation by basic packaging
- Safe storage methods
- Microbial shelf-life quality criteria for foods
- Bioindicators for food quality.

The principle of maintaining a long shelf life is central to the regulatory aspect of food and drugs, as well as other products. The length of shelf life will vary according to the raw materials used, the nature of their processing, the application of preservatives, and microbial contamination. In general, foods with a long shelf life are safer as they contain little or no water, which is what microorganisms need to grow (Nirupama et al., 2010). Therefore, the longer a food item stays on the store shelf, the less likely it is that it will be contaminated with spoilage microorganisms. This is why it's important to maintain foods on the shelves for the longest possible period before use.

Figure 4.3: Organizations normally sell their goods to retailers making them the most crucial pillars in the regulation of shelf life.

If a food has a short shelf-life, the organization selling the products will most likely sell them to the stores and not directly to the consumer. Longer shelf-life foods can be sold by many different outlets, including supermarkets. Regulations on the length of time that food may remain at room temperature are based on assigned risk categories. The assigned risk category is based on how susceptible food is to bacterial growth that can cause illness. For example, an opened beverage will have a shorter shelf life than a raw vegetable that has been washed thoroughly.

While clients often think of shelf life as relating only to the time during which products are sold, the term "shelf life" is legally defined in several ways. Adequate shelf life can be critical for several reasons, including economic ones. One can rack up significant losses if an expired product is sold, or if a product becomes contaminated because it is not stored appropriately (El-Anany et al., 2009). Any firm planning to produce, handle, store, transport, or sell products should have a grasp on the general principles relating to shelf life and Federal regulations related to it.

Regulation of the shelf life of products is a major concern for product manufacturers. In most cases, regulations stipulate how long a product can be stored in particular conditions, after which it should either be discarded or reworked, respectively. Monitor the shelf life of certain food products in a modern, highly specialized, and globalized food production process, which is characterized by an increasing number of actors, several scientific approaches have been used. The traditional one is based on microbiological analysis: after the sensory properties of freshness are no longer apparent, spoilage is identified through microscopic analysis of the product. However, the physical-chemical analysis of food products (from their constitution to the degradation processes) has increased in recent years due to new techniques like mass spectrometry or infra-red spectroscopy and their cheap commercialization. These are used to identify chemical markers of deterioration that allow a food product to be defined as spoiled (which allows for judicial proceedings) or to determine at what point a product is no longer suitable for human consumption.

4.2 DETERMINATION BY THE POTENTIAL HAZARDS TO HUMANS AND ANIMALS

When deciding on the shelf life, it is always important to consider the potential hazards to both human and animal health when determining how long a food item should be kept before consumption. Regulated and nonregulated food products must be stored at temperatures designed to retard the growth of spoilage microorganisms, enzymes, and plant and animal pests. Spoilage begins to occur when harmful microorganisms begin to grow and multiply. Eating foods that have an expired shelf life is not recommended. However, it's up to you to decide whether you are willing to consume these items after the recommended date.

The United States for instance has a tradition of providing advice on how long a product should be stored. This is usually in the form of several days or months that it can be stored safely and sold. Sometimes this information is just 1 number, but more often it is given as a range (Nagi et al., 2012). In almost all cases, the reason for providing this information is to protect consumer health as well as food safety. With agencies working together to ensure that food is safe for consumption through the use of scientific findings, food can continue to be sold without causing harm or illness to humans or animals.

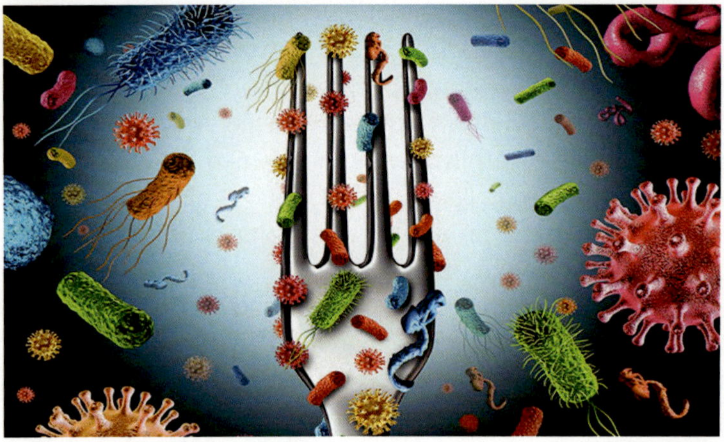

Figure 4.4: Foodborne diseases are some of the risks that relate to hazards to humans.

The use-by date and shelf life of food: the principle for determining is formulated with using humans and animals in mind, it does not focus on how the production process is flawed. Food companies and FDA need to be more aware of potential hazards to humans and animals. This is the basis of considerations on those factors which may have the most significant effect on public health and well-being, including the hazards of microbial contamination, chemical contaminants, and physical deterioration.

In a case example in the united states, foodborne outbreaks have been reduced in recent years, data shows that the number of outbreaks associated with high-acid foods (pH 4.6 or below) has increased. According to studies by the US Food and Drug Administration, low-acid foods are responsible for more than half of all foodborne illnesses associated with long-duration storage (greater than a week). Foods with pH values below 4.6 pose minimal risk. This then creates a basis for determining shelf life (Maertens et al., 2012).

Based on this principle, a regulation is proposed on determining the shelf life of food when potential hazards arise. These hazards include; microbiological hazards, toxic hazards, and radiological hazards. Chemically they represent substances that are toxic in foods. These substances have been recognized as contaminants in foods that have negative effects on human health or animal health. Moreover, this relationship between collection data and risk factors has been documented, making it an indispensable guide for food science and technology professionals and other researchers.

This principle is a guiding basis for the maximum shelf life. The maximum shelf life determines what products shall be allowed to remain on the market. Classification of the danger toward man and animal is made according to the so-called "maximum shelf life principle", which means that the shelf life limit cannot exceed the period that a product may present hazards for man or animal in case of normal use. Generally, a shelf-life determination is based on physicochemical indicators such as appearance changes, flavor and odor changes, and chemical composition changes in food. It has been reported that food-based pathogens induce a shortening of the shelf life.

It is argued that by using scientific knowledge about the complex factors affecting harmfulness, it should be possible in the future to base regulations more closely on the actual risk involved. The advantages of such an approach are compared with those of conventional systems involving shelf-life adjustment or prescriptive storage conditions (Manzocco, 2016). This principle eliminates the need to predict the cumulative effects of (minor) single attributes and reduces shelf life determination to an assessment of potential hazards---of a functional, microbiological, or unnecessary chemical nature. The evaluation criterion requires the attribute to be potentially harmful as opposed to being likely harmful.

The meat and meat products industry has been trying to implement the "benefit" approach, based on the determination of the changes in shelf life, in establishing the shelf life for processed food products. Usually, the changes are determined as a result of microbiological changes and/or chemical changes. Even though this approach meets the safety requirements and is specific to each product, it results in shelf life values that may be too short in some cases.

In the United States, the work of determining potential hazards is placed on the Hazard Analysis Critical Control Point (HACCP). The U.S. FDA's Advisory Committee on the Microbiological Criteria for Foods (ACMCF) set out to develop a worldwide consensus on food safety control based on the concepts of Hazard Analysis Critical Control Point (HACCP). To achieve this goal, a comprehensive examination of the microbiological criteria was required. The ACMCF set about developing a rational, scientific method for determining shelf life for foods. A HACCP procedure was developed using a toxicity-based paradigm. Data regarding the environmental growth requirements of pathogens and their resulting illness caused by their ingestion were used to determine the growth potential associated with temperatures that a product may be exposed to during its distribution and storage life (Fan et al., 2009). The approach applied to food is similar to those used to establish the Minimum Required Temperature (MRT), an important analytical tool that allows processors to ensure safe shelf life in refrigerated foods. Such qualitative analyses have inherent limitations but they are effective indicators of many processes.

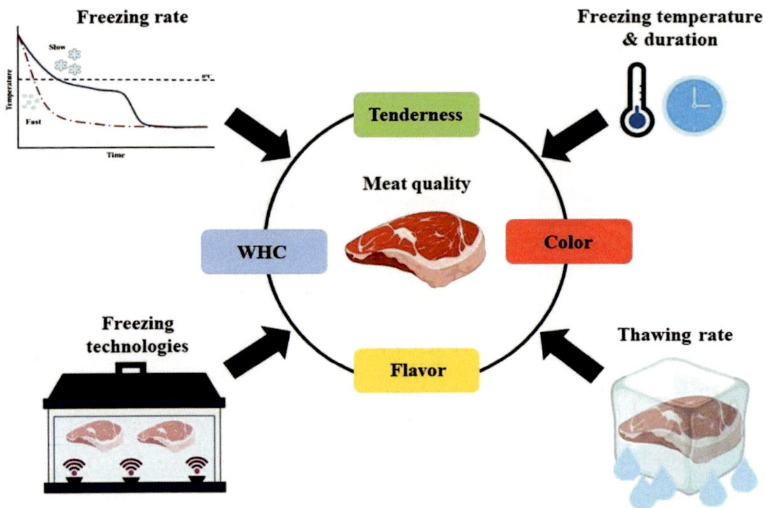

Figure 4.5: Contemplating the minimum required temperature is one of the most crucial analytical tools.

(Source: Dang DS, Bastarrachea LJ, et, al., 2021)

Consumer interests are best protected through prevention rather than intervention. Prevention requires that the food industry and the government agencies responsible for ensuring the safety of food in commerce take measures to ensure that potentially hazardous foods do not enter the marketplace. An example of a potential hazard ever recorded is the *Listeria monocytogenes* foodborne pathogen. It is the second most prevalent after Salmonella, which has claimed the most lives (approximately 2000 deaths/year in the United States alone) among those caused by foodborne illnesses. While the general public is accustomed to overhearing Salmonella outbreaks, there is no doubt that *L. monocytogenes* will also deserve the same concern (Galić et al., 2009). Interestingly, this microorganism requires less amount of bacterial cells to be injurious than Salmonella, which implies that it should not be so easily disregarded. Moreover, a huge number of foods may become contaminated with *L. monocytogenes* as this organism seems capable of surviving under multiple conditions (e.g., in foods and on a range of processing equipment). It thus prevents the guide from monitoring shelf life regulations.

4.3 STABILITY OF PRODUCT

The stability of a product as a principle in regulating its shelf life is the primary merit of various legislations across the world. A product under different storage conditions, changes in molecular structure, molecular mass, molecular weight distribution, and activity of polymer will lead to changes in its properties. The shelf life of a process shall be regulated by the stability of the product, using biochemical or chemical analysis or other proven methods to ensure that the critical microbiological quality parameter is not exceeded.

Figure 4.6: The stability of a food product is determined by the condition.

Taking the case study of a meatball product, Raw materials (rabbit meat and offal, water, seasoning mix, thickening agent, salt) were carefully selected according to their microbiological quality (*Salmonella* spp., *Escherichia coli*, *Pseudomonas* spp. aerobic mesophiles, Enterobacteriaceae), by European and national criteria. They were comminuted and well mixed by adding 45%

wt of water (raw materials/water ratio: 1/0.45). The meatballs were cooked under an isothermal regime at 60 °C for 2 h. The analyses performed during the storage period (1-25 days) showed that both cell growth characteristics and chemical changes pointed to a long shelf life (GÁMbaro et al., 2006).

In the medical industry, stability is a chemical property related to a molecule's tendency to decompose into simpler molecules. By understanding what causes medicine to have instability and how to counter unstable reactions, pharmacists can create medications with the longest possible shelf life. Similarly, the shelf life of food or beverage products depends on many factors, but the key component for the shelf life of a product is the stability of each component. Stabilization products that improve flavor and prolong shelf life are essential to any packaging operation.

The stability of a product is directly triggered in the manufacturing stage. When the first product is made at the end of the manufacturing line, it has been manufactured and has a shelf life of its own. All subsequent products may be made from that same lot without violating said principles for stability as long as each unused batch of active ingredient(s), each intermediate, and all finished products so made are identified. In addition, manufacturing steps that increase the general stability of the product by reducing microbial contamination should be utilized, including adherence to sanitation procedures such as those requiring personnel to wash their hands, use clean coverings for their clothing, and use dedicated equipment for specific purposes.

As food processing technology advances, so does our knowledge of how to keep food safe longer. We now know that food safety depends not only on when it is harvested but also on its storage temperature. In other words, the basic idea behind food safety regulations is not just to set expiration dates after which foods are dangerous; rather, shelf-stable products are expected to remain safe for much longer than the printed date (Manzocco et al., 2010). They should remain safe for as long as their physical barriers – their shell or wrapping – hold out. When designing or developing products, or when formulating or reformulating products, it is always desirable to maintain the physical, chemical, and microbiological stability of products for longer shelf life.

A shelf-stable product does not need refrigeration after opening. Benefits include convenience for the consumer and less energy for product transport and storage. Criteria for evaluating stability include sensory attributes, microbial safety, chemical safety, the efficacy of any preservative system, protection of food from environmental damage, and finally oxidation. A product is said to be stable if it maintains its quality and physical characteristics, and therefore, can be stored for an extended period without changes in its properties. Understanding the role that storage conditions and product stability play in maintaining product quality throughout its shelf life can assist one in making prudent decisions toward minimum inventory levels.

4.4 POST-TREATMENT PROCESSES

Around the world, more and more consumers are demanding a variety of high-quality foods and drinks. Post-treatment processes have therefore become an important pillar in the effort to regulate shelf life. When fruit and vegetables are treated properly after harvesting, they lose much less nutrition than when transported without treatment. The result is increased freshness and flavor in the finished product, with benefits for both producer and consumer.

The post-treatment processes, in terms of storage temperature, humidity, and presence or absence of oxygen, are all connected with the change of oxidation status of food materials during the storage process. These processes will play an important role in regulating shelf life. In the meat industry, the addition of post-treatment processes optimized to reduce the number of contaminants already on fresh meat is a significant contributor to reducing the microbial load and maintaining a high-quality product. This lowers the microbial load at both processing plants and retail markets (Galovičová et al., 2021). When you order a steak medium-rare, the chef has a variety of ways to heat it. Over time, meat can be heated to cooking temperature in many different ways. As a consequence, the color change and shrinkage of the meat will vary substantially from one recipe to another. This variation in appearance explains why a medium-rare steak may not look exactly like its picture on a menu.

Post-treatment processes include; blanching, pasteurization, high-pressure processing, pulsed electric field processing, and radio frequency treatment. There is a need for an efficient and cost-effective alternative to controlling shelf life in dehydrated foods. The current practice of pasteurizing products after dehydration and inner vacuum packaging is an expensive alternative that compromises the food's sensory attributes while adding little value in terms of safety and stability.

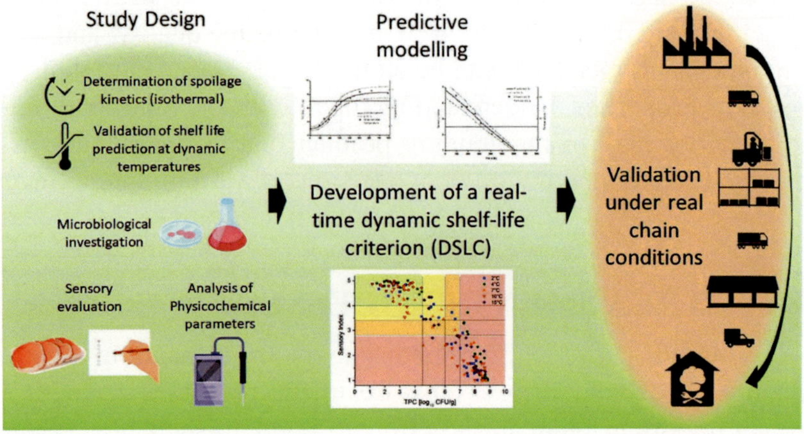

Figure 4.7: Pre & post-treatment processes determine the shelf life of food products.

(Source: Albrecht A, Mittler M, Hebel M, et, al., 2021).

Just as certain foods have their particular aromas or tastes, and others are always crunchy or sugary, foods have shelf lives of varying lengths. All foods deteriorate with time: some do so to a greater extent than others. Unwanted changes that occur in food products generally come from contact with air (oxidation), contact with water (ocean influx), or enzymes naturally present in the product itself (Martins et al., 2008). Increasingly strict legislation caters to more efficient and safer food preservation methods and all the while new challenges come up – environmental protection, health, and safety issues such as *E. coli* bacteria, hydrogen sulfide gas produced by the anaerobic fermentation of organic debris, energy-saving–that are forcing food processors to find optimized solutions.

Halophilic and halotolerant bacteria have successfully been used to treat foodstuffs and make them suitable for preservation without refrigeration. The shelf life of a product, or the period during which it remains in good condition and fit for consumption, is determined by the potential of the microbial population to develop, multiply and spread. Traditionally, to extend the shelf life of a food product, chemical preservatives have been used which could be toxic.

The shelf life of certain foods is relatively short and highly variable. To deal with this problem, a scientific assessment of the food's chemical composition, knowledge of how this changes during processing, an understanding of the relationship between chemical structure and stability, and robust models to predict changes in food properties are required. The current state-of-the-art requires a combination of scientific expertise with out-of-the-box thinking to help identify key factors controlling the chemical composition and predict changes in food properties during processing that influence the shelf life of different high-value foods.

4.5 LABELING INDICATING

The gist of the law stipulates that all goods should carry "labeling indicating as a principle in regulating shelf life". The primary objective of shelf-life determination is to ensure that acceptable product quality is attained up to the end of the designated shelf life. In the United States and Europe, use by or best before dates are required on foods that have a reasonably long shelf life. In most countries, this information is required on labels (Lindström et al., 2006). In some countries, including Canada, Australia, and New Zealand, date labeling is not mandatory; however, manufacturers may voluntarily label their pre-packaged products with a "best before", "baked on", or "packed on" date. And in most cases where these dates are labeled, there is no requirement to include detailed storage instructions.

Figure 4.8: An example of labeling indicators on a product.

The Code of Practice for Use-by date, an expiry or use-by date on pre-packaged food products is a definite indication by the producer or supplier that, if the product is stored correctly, it should remain wholesome and edible until a specified time. With the widespread use of food additives (especially preservatives), many consumers have become concerned about the possible effects of food additives on human health. Therefore, there is an increasingly strong tendency to use natural and safe food ingredients or to postpone the purchase of goods with a relatively long shelf life. Emphasizing the principle of "labeling indicating as a principle in regulating shelf life" will help promote consumption patterns of goods with a shorter shelf life and promote some industries to develop products with a shorter shelf life (Liu et al., 2018).

The international Packaging Convention defines the "best before" date as a means for indicating that a food's quality will fall below a particular standard by that date. Labels don't just look pretty, they can also help you understand the shelf life of foods in your fridge, freezer, and pantry. It all comes down to two numbers – one on the label indicates when the food was made; the next number tells you how long it will last after purchase. It allows special attention to be focused on certain types of food (Kanatt, 2020). It may be used in raw and prepared foods (e.g. fresh, chilled, and frozen foods). The regulations involve many different variables, including the quality and type of food product and its packaging, as well as storage temperature and atmospheric environment.

4.6 RELEVANT LEGISLATION CONCERNING TRACEABILITY

In food law, the term traceability encompasses not only the possibility of identifying a food at any given stage in the supply chain; it also indicates that there is a legal obligation to keep records and preserve the records for a specified period. Food is often still rather tasteless long after its sell-by date. In an elaborate example, a mandatory traceability system has been introduced in the EU on 1 January 2018 to enable food waste to be reduced, especially in supermarkets and grocery stores. Shelf-life information can now be rendered more precise by the requirement for suppliers to notify their customers of any deviations. In addition, the legislation stipulates labeling requirements: The listed sell-by date must not be misleading, but also foreseeable for both supplier and consumer.

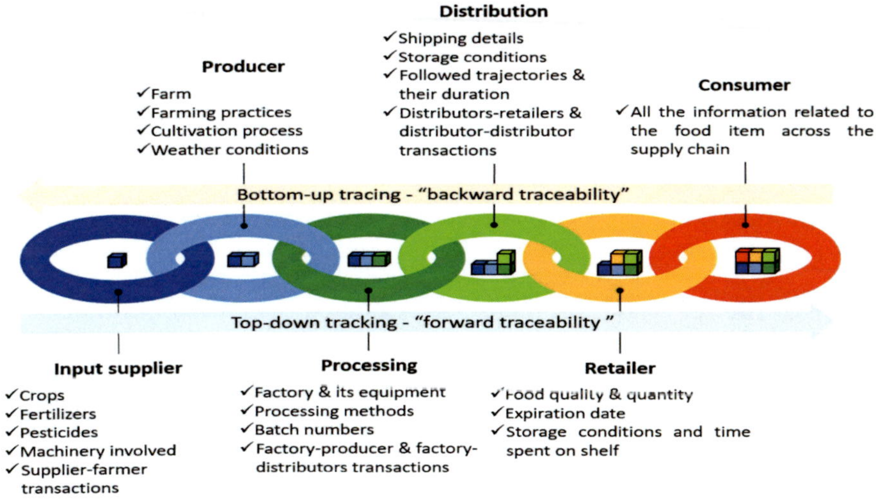

Figure 4.9: Food product traceability is now among the trends in regulating shelf-life.

(Source: Tagarakis AC, Benos L, Kateris D, et, al.,2021)

The traceability system is based on a series of control points that the product moves through and which must be recorded. This record can then be used to determine the movement of a product in either the forward or backward direction. Two types of records are used within this system; Activity Records, and Security Records (primary records). An Activity Record provides information regarding when and where an individual activity was conducted relevant to a food case. On the other hand, a Security Record provides information regarding when and where a 'specific' food case was moved relevant to its food safety status (Kuijt, 2009).

In addition to good manufacturing practice, this principle requires that a manufacturer establish and maintain a procedure for tracing allergenic substances in its products to their sources of raw materials or further processing. Such a procedure must be documented and subject to control by the quality control personnel of the company regularly. The main purpose of such procedures is to provide information/evidence when there are allegations of potential direct or indirect cross-contamination of one or several ingredients, or production lot.

Under the traceability system, there is a common requirement for the complete traceability of foodstuffs between the various links in the food production chain. Food business operators are obliged to establish traceability systems to ensure that food can be traced back and forward. Similarly, labeling must contain sufficient information to allow tracing within the supply chain. Hereby, if a problem occurs, affected products can be readily identified and removed from the food chain.

4.7 MICROBIOLOGICAL AND PHYSICO-CHEMICAL PARAMETERS

The shelf life of food products is regulated by microbiological and physicochemical parameters. The consumer demand for high-quality food has led to increased research on new preservatives, antioxidants, and packaging systems, leading to the extension of the shelf life of a product. The shelf life of a food product is the period during which the product can maintain its desired quality and nutritional characteristics. To guarantee the safe consumption of a food product and to provide high-quality products for consumers, it is necessary to develop a rational shelf-life control method based on changes in microbiological, physical, and chemical indicators that are closely related to food deterioration and quality in foods (Kendall & Dimond, 2007). This method is suitable for controlling all kinds of food products with long shelf life and guaranteed good stability.

A combination of microbiological and Physico-chemical parameters as a principle in regulating shelf life can be used to replace traditional approaches according to which shelf life is based on one parameter. The ratio of moisture content, microbial load, pH, and chloride has a direct effect on the shelf life. Most of the current methods to predict shelf life rely on predicting microbial growth. However, as food systems become more complex with high processing techniques, there is an evident need for improved shelf-life prediction methods that are led by physicochemical parameters instead of microbiological studies.

Microbial growth and metabolic activity are the major important factors that affect the shelf life of most food products including dried pomegranate seeds. The parameters that can be used to predict microbial growth and metabolic activity include pH, moisture content, microbial count, peroxide value, and biochemical changes due to microbial respiration. The proposed regulation of food shelf life based on microbiological and Physico-chemical parameters is a breakthrough in the food industry, replacing vague assumptions and tacit knowledge of it with a precise assessment of the shelf life. All data obtained on this principle has potential practical application in determining product shelf life in foods, which is mostly based on experience with little or no scientific basis. It enables further improvement in the quality uniformity of production by taking into account the storage conditions.

Food is perishable but also has a relatively long shelf-life, thus a mechanism is required to ensure food safety. It can be inferred that microbial quality is the master for preventing food spoilage and extending shelf-life, as demonstrated by many previous studies. The shelf life of different foods, such as yogurt, fresh cheese, butter, and mozzarella, depends essentially on the correct manufacturing process, or on the realization of the essential quality control measures. For example temperature during storage, pH value, water activity (aw), and keeping salt content as low as possible are just a few key factors for optimal shelf life. The microbiological and Physico-chemical parameters act independently of the product (Howey, 2015). Underlying mechanisms can be a change in the redox potential and action of enzymes which lead to browning reactions or rancidity. However, it is not only the production conditions that are decisive for the shelf life. Every step from processing through packaging until consumption may adversely affect the durability of a foodstuff.

The Hurdles technique, based on the principle of regulation of Physico-chemical parameters (temperature, water activity, pH), can offer an alternative. This technique is an optimization tool that allows one to determine acceptable values of these parameters to prevent the growth of pathogenic microorganisms and extend the shelf life of products. There is no shortage of new technologies, among which we will mention ohmic heating, pulsed electric fields, and high hydrostatic pressure. They help to guarantee the microbiological stability of food items without causing any significant changes in their nutritional composition or sensory value. The study investigated how different parameters such as pH and aw affect the shelf life of four dairy products: long-life milk 3.2%, sterilized cream, fermented milk drink, and cream cheese.

4.8 HYGIENIC PRODUCTION AND PRESERVATION BY BASIC PACKAGING

A hygienic production and preservation by basic packaging as a principle in regulating shelf life can get rid of all kinds of bacteria, disease-causing microorganisms, viruses, and mold. A cold chain that includes HTIC controlled temperature control technology at the beginning and end of long-distance transportation will be able to raise the requirements for quality inspection, minimize customer complaints about expired products and directly improve customer satisfaction (Hedegaard & Skibsted, 2013). Basic packaging is the hygienic production and preservation of the product by a combination of packaging processes and technical measures. Basic packaging products are packaged on the principle that they fulfill their characteristics and functions if they are stored in the appropriate conditions for storage and use.

Figure 4.10: Food products ought to be preserved hygienically.

It is the appropriate category for packages in which the main function is to physically protect the contents from damaging influences (e.g. mechanical, electrical, and/or thermal shock, pressure and/or humidity, oxygen) by denial of direct contact with the contents with their environment or to provide them with the right conditions for useful content (protection against oxidation of fats and oils, preservation of grain quality). Indeed, with the development of modern society and the improvement of living standards, our lives are increasingly complex. Hygienic production and preservation greatly improve the quality of agricultural and aquatic products, ensuring their quality, increasing storage stability, and extending shelf life.

Today consumers are being confronted with new challenges and choices in the food market. To satisfy consumers' demands, there is a need to be creative and devise a concept that integrates functions such as preservation, freshness, and hygienic production into a total package to provide safe and fresh foods. A mold is one of the necessary factors for the pathogenesis of foodborne diseases and a great threat to food and health. Inappropriate packaging can reduce the growth of molds and bacteria in foods (Goldman et al., 2002). Infectious microorganisms, such as bacteria and mold, can attach to plastic materials and penetrate the tiny pores. Once wet, it will be difficult to dry out without deformation. However, with basic packaging and proper handling, it can reduce the risk of foodborne diseases caused by mold.

4.9 SAFE STORAGE METHODS

The life expectancy of foods is a subject of some debate. The quality and safety of the food can be affected by the storage conditions. It helps reduce the risk of foodborne illness by storing food in ways that will reduce the loss of nutrients and flavor, or otherwise improve shelf life. Prolonged storage can lead to the growth of microorganisms and the formation of mold and off-odors. The shelf life may be significantly reduced as a result. In principle, food in storage should be preserved in such a way that it is protected from unfavorable conditions, such as extremes of temperature or humidity, foraging animals and vermin, fading of flavors and colors, etc. Using the appropriate storage conditions is, therefore, an important part of ensuring quality and prolonging shelf life.

Figure 4.11: It is not only refrigeration but a subtle way of storing food that assures shelf life.

When it comes to the storage of food, most people think only of refrigeration. However, other methods of food preservation and storage can be just as effective and safe as refrigeration. In many cases, a safe shelf-life can be extended by using these alternative methods. Such methods include cold, dry and frozen storage. Many people believe that bacteria only multiply in stressful conditions, therefore they thought food products will keep longer at low temperatures. However, this is not true. Salmonella, Listeria, and other harmful microorganisms can still grow in the refrigerator up until the expiration date on the product label (Hosseini & Jafari, 2020).

Understanding the principles of safe storage and the methods used to control microbial growth and food quality in refrigerated food products helps retailers achieve the maximum life from a product. Labels containing instructions for proper storage methods help minimize the risk of medicated meat and poultry products becoming unsafe due to microbial growth. The instructions serve as a guide for maintaining the product's shelf life through proper cold chain management. Food poisoning and spoilage are caused by unsafe storage conditions. Safe storage conditions can help to limit the probability of contamination and hence, food poisoning and spoilage.

Every day, people from all over the world make critical decisions about food safety, from where to buy and how to prepare it, to whether or not it is safe to eat after it has spoiled. Many of these choices are based on the amount of time the food has been stored and the conditions under which it has been stored. Food can be preserved in many ways, from freezing and cooking to drying and salting. Throughout human history, food spoilage has often occurred as a result of improper handling, packaging, shipping, and storage methods that caused food quality to be impacted by temperature and humidity changes in the surrounding environment.

But what is the essence of proper storage? Food spoilage is mostly caused by the growth of spoilage microbes such as yeasts, molds, aerobic bacteria and other microorganisms. These microbes may be present in the product itself, or they may enter the material from outside sources. They compete with yeast and mold propagules on the surface of foods for available nutrients. The first microbial cells to reach sufficient population size cause food spoilage by reducing the pH below 4.6 or by using sorbitol or fructose, which are less inhibitory than glucose, as their carbon source.

4.10 MICROBIAL SHELF-LIFE QUALITY CRITERIA FOR FOODS

Many new food products are designed to slowly release flavors, textures and scents over long periods of time. That is why there is a growing need for microbial shelf-life quality criteria for foods as a principle in regulating shelf life and the techniques for calculating microbial upper limits for shelf life for foods. Microbial shelf life criteria for foods as a principle in regulating shelf life is a prerequisite toward extending the shelf life of foods. Industrialized countries struggle to encourage food producers to comply with stringent food safety regulations (He et al., 2018). The majority of food-borne illness cases are caused by foods that are beyond the shelf-life date. The food manufacturer's cost of compliance is high because proper training is needed to ensure employees do not make mistakes in tracking lots and series of dates while filling packaging materials, recording dates, implementing proper storage and handling controls, as well as the implementing of accurate test methods.

This paper proposes that microbial shelf-life is a principle for the regulation of food safety and quality in developed countries. Food microbial shelf-life quality criteria are proposed to fulfill three functions: (1) guarantee product safety, (2) provide information to consumers about the relative safety of various products by communicating product bacterial load, and (3) allow for more consistent use of enforcement action, thereby leading to better compliance with legislation. A shelf life criterion was recently proposed which equates microbial shelf life with the estimated number of days left until an environmental pathogen will perform a certain detrimental effect on the product. Before this, shelf life was defined by expert judgment based on experience and mostly untested scientific principles.

The two main principles driving this novel definition for determining the shelf life of food, either as a single product or a combination of products are: 1) The demand for high-quality food products from consumers and producers. Food that is fit for consumption while maintaining appearance, flavor, and taste throughout its shelf life to prevent food waste generation and 2) the precautionary principle to protect public health by not allowing contaminated foodstuff on markets or in kitchens. These two principles have been combined into one basic principle - showing how the potential benefits of this concept are much larger than its potential costs (Zhukov et al., 2018).

Microbial shelf-life is the time by which a food will be unacceptable for consumption due to microbiological spoilage microorganisms growing. Microbial growth occurs through 4 different steps, the 4 stages of growth are the lag phase, log phase, stationary phase, and death phase. A food is declared within its shelf life if it passes inspection during each stage. Sensory, microbiological, and chemical attributes can currently be quite powerful in determining the shelf-life of a food product.

The problem is that shelf-life is considered independently from the time after the last point of sale. Shelf-life is directly related to the time after the last point of sale, making it an incorrect protocol for food recall purposes: Cephalosporium (cep) and Aspergillus (asp) fungi are sensitive 0r 2, 3 period survival times after purchase and therefore represent a leading indicator of quick death or spoilage, thus when they're present at low levels they tell you that your unit has spoiled. This microbial information can then be used as the substrate for shelf-life determination using traditional sensory techniques.

Many consumers today demand foods that have a longer shelf life. In some cases, this is caused by the increase in demand for processed foods or meats that need to stay fresh even if they travel long distances to the consumer. Some of the ingredients used in food packaging are designed to inhibit microbial contamination of the food. The inhibition needs to be effective only during the lifetime of the product and often this will be much less than two years.

4.11 BIO INDICATORS FOR FOOD QUALITY

Bioindicators for food quality as a principle in regulating shelf life (B-Fresh) is a unique blend of natural testing microorganisms that serve as an indicator of freshness, preserving the quality and safety of food. B-Fresh can be used to control the biological parameters in refrigerated storage including temperature, humidity, and oxygen and water activity. These parameters are typically altered during transportation or distribution resulting in loss of flavor, nutritional components, change in color and texture, increase production costs and possible food recalls due to deterioration (Weinbreck et al., 2010).

The introduction of food-quality indicators and the consequent regulation of shelf life by these indicators is accepted by Singapore's food and drug industry. The use of food-quality indicators has been adopted in Denmark and Sweden. In Sweden, offal, pork, beef, lamb, and chicken are monitored using bioindicators. The shelf-life of food is defined as the period between manufacture or harvest and the point where the product reaches unacceptable quality. Bioindicators are good predictors that are put in foods to count microorganisms and decide shelf life. Bio indicators may use substances such as urease, adenylate cyclase, catalase, galactose oxidase, etc.

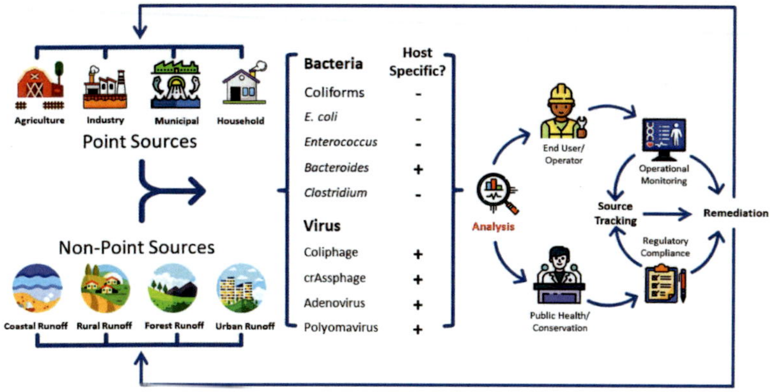

Figure 4.12: Sample bio-indicators of food products.

(Source: Li E, Saleem F, Edge TA. et, al.,2021)

I have a new approach to product quality control. The foods which have turned bad cannot be eaten no matter how -long the shelf life is. It has a big problem in the food industry to solve it to give a scientific reason and make a proper assessment, the author, by using bio-indicators through planar laser-induced fluorescence (PLIF) and luminescence (LIF), verified that bio-labeling can effectively identify food quality and life status of food, thereby laying a foundation for developing more effective, convenient and reliable food quality regulation system (Villalobos-Delgado et al., 2019).

Significant quantities of pollutants (including products, explosives, and drugs) have been found in food samples. Bioindicators are necessary to verify the quality of food raw materials, and products and to identify sources of contamination in the food chain for human, veterinary, and livestock foods.

4.12 REQUIREMENTS FOR AN EFFECTIVE SHELF-LIFE PROGRAM

Consumers are often confused about the shelf life of their food, which means that effective communication of the shelf life of your products is vital to successful sales, customer satisfaction, and protect your brand. Dehydration and packaging technologies can preserve the freshness of your product for up to two years after production if one has an effective shelf-life program.

This ought to; the product to be stable and deliverable, the process must meet your customer expectations, the quality assurance program, and all other requirements for stabilization and maintenance of the product's intended shelf-life. Documentation and test results of all tests performed on a product being submitted for storage can be used to justify that the product has met all these requirements. Core products need to be carefully selected. The shelf-life of the product is tested and a point is found when quality starts to decline. The shelf-life must be long enough to ensure that most customers use the product before the quality is too poor. If the shelf-life is too short then shops have too much waste in terms of unsold products. If they are too long, then shoppers will lose confidence in the shop due to poor quality produce. Specific shelf-life requirements will demand various combinations of monitoring, testing, and decision analysis.

Strict maintenance and control of the inventory are essential for products that have a limited shelf-life. This is achieved through the use of Item Control Codes which control parameters such as the: creation of products, receipt into stock, issue to users, expiry date and shelf life, minimum/maximum stock levels/reorder points, etc. The introduction of stringent economic quantity policies must be managed properly to ensure that sufficient requirements are in place to carry out planned maintenance activities and to replace obsolete spares (Wrona & Nerín, 2020).

A shelf-life program should be conducted in four overlapping phases: Phase 1: Product Identification, Phase 2: Product Evaluation, Phase 3: Product Testing and Validation, and Phase 4: Implementation (including ongoing monitoring and corrective action). The shelf-life program consists of well-documented and controlled procedures to ensure that the product's identity, strength, quality, purity, and potency are maintained during its expected shelf life. This program ensures that the product meets specifications for stability requirements. A plan is developed by operations/QA/quality to monitor stability and establish lot expiration dates based on stability data from assigned studies. The shelf-life plan does not supersede any regulatory requirements for the establishment of the end-of-expiration date.

4.13 ENACTING REGULATIONS RELATING TO STORAGE AND LABELING

Agencies such as the European Union Food Information for Consumers Regulation 1169/2011 (FIC) lays out the current requirements for shelf-life and use-by dates on food products. The FIC is relevant to food sold to consumers, including food sold at catering establishments. Food products must meet the definition of a foodstuff as defined in Article 2 of Regulation

(EU) No 1169/2011 and must comply with conditions on ing labeling and marketing rules. The safety of all foods within the EU is subject to a general presumption that any substance which might endanger human health is banned from use as an ingredient in the food (the so-called 'general safety principle').

4.14 PRE-TREATMENT FOR SHELF LIFE IMPROVEMENT

There's a way to help your products stay fresh on the shelf: it's called shelf life pre-treatment. A special process that helps slow down the rate at which food loses its flavor and texture, it can be applied to products while they're already in production. This innovative formula can help food firms accelerate the industrial path to success by allowing them to shelf-stabilize certain edible products. While it is ideal for use in fresh, frozen, or shelf-stable applications, this unique solution is also great food safety and convenience product for the food service industry.

The pre-treatment for shelf-life improvement raises the water activity level in prepared meats, thereby slowing the effect of moisture loss and consequent weight loss while maintaining sensory and nutritional qualities. Prevents color, texture, and taste changes in ready-to-eat fresh meats and fish by combining a high concentration of salt (>11%) with sucrose and methylcellulose, forming a protective coating on cut surfaces of shelf-stable meats and fish during storage (Winterhalder et al., 2015).

4.15 THE PROVISION OF FOOD INFORMATION TO CONSUMERS

Consumers expect accurate information about the food they buy and consume, and they rely on the food industry to provide this. In this report, we discuss how the food industry can meet its expectations around accuracy, consistency, and completeness of product information communicated to consumers. The U.S. food supply is among the safest and most bountiful in the world. In an increasingly globalized economy, it is important to know when food products originated, where they were grown and how they were processed. The provision of such information to consumers has generated a response across all sectors of the domestic and global food industries, as well as government regulation on a national level. Simply put, this diverse market is working behind the scenes to ensure a safe, abundant, and highly efficient food system for Americans.

Maintaining a food product's quality after it leaves the processing facility is challenging. The possibility of temperature abuse, whether intentional or accidental, makes it difficult to determine how to send the product to market for the maximum length of time. By providing consumers with information about storage and shelf life, firms can help them extend product shelf life and reduce waste. Food date labels, or expiration dates, are designed to help consumers protect their families from products that might make them sick. Yet according to recent surveys, 87% of consumers have no idea when foods are spoiled.

4.16 CONTEMPLATING THE PERIOD OF USE AND THE PERIOD OF VALIDITY

As the core component of quality management, product shelf life is a common issue for all consumers. In a long-term consumption process, mastering the product's shelf life is equivalent to the "service life" of goods storage, transportation, and consumption. Usually, people automatically associate the word "shelf life" with food or daily necessities. Ingredients are added to a variety of cosmetics products to ensure their quality. The purpose of adding various ingredients is to obtain antibacterial action, and prevent products from changing color or deteriorating in quality during processing, distribution, and storage. Highly regulated shelf-life statements (eg, "use by" and "best before" contribute to food waste as many consumers and retailers are not aware that these labels do not indicate safety, but rather quality.

Many factors need to be considered in managing the shelf life, such as the production line production capacity, the product life cycle, actual sales performance, etc. For example, the shelf life can be adjusted according to the seasonal changes in sales volume, to satisfy sales needs and keep products fresh (Wibowo et al., 2018). Under the ing labeling rule of practice, the validity period (also known as the shelf-life) of a product is only regulated by the manufacturer or supplier and is not controlled by any government department. The expiration date is set according to scientific research or experiments and may be related to certain conditions such as storage methods, temperatures, etc.

The consumer must not be in error, even if the incorrect expiration date is due to incorrect translation and translation of product information or packaging labels, or if he/she is not familiar with the legal language in the country of purchase.

The food industry and health authorities throughout the globe are clamoring for verified methods for determining the shelf-life of food products. When food is stored, it undergoes structural, compositional, and physical changes that affect its shelf life before it may be consumed. Physico-chemical, environmental, and microbiological variables all have a part in causing these alterations, which include spoilage bacteria, which play a significant role. Sensorial degradation is "translated" into these changes at a certain moment in time. Despite posing a health risk to consumers, food-borne bacteria do not affect alterations in sensory perception.

The environment in which pathogenic bacteria may develop and/or create toxins may be affected by the opening of the food item (i.e. extrinsic, intrinsic, or implicit factors). Extrinsic elements (such as the makeup of the environment) are likely to alter the most once the package is opened. In the absence of vacuum or modified atmosphere packaging (MAP), the pathogens in the food might be predicted to modify their growth behavior (often by increasing their development capability/rate). After opening the packages, pathogen development may be influenced by intrinsic (such as aw or pH) and implicit (such as competing microbiota) variables.

CHAPTER 5

FACTORS INFLUENCING THE SHELF LIFE OF FOOD PRODUCTS

CONTENTS

5.1 INTRODUCTION

Packaging is an important part of the food manufacturing and distribution process. Even while food preservation is the most essential purpose of packing, the food producer must be cognizant of a variety of others. Packaging is threatened by insects, rodents, and diseases. If oxygen and water vapor are allowed to enter the packaging, they will harm the food.

When it comes to a product's shelf life, packaging might be a limiting issue. Both contaminated food and contaminated packaging can move into one another, resulting in contamination. Consumers, merchants, distributors, manufacturers, and farmers all have slightly varied perspectives on shelf life, depending on what is most relevant and significant to them. Items must be safe and of high quality to keep customers satisfied. Consumers usually search for the product on the shelf with the longest shelf life, since this is supposed to signify freshness (Uebele et al., 2013).

Food has a "shelf life" when stored in a refrigerator or freezer as long as all storage standards are satisfied. A product's shelf life begins the time it is prepared or made. Its shelf life is impacted by a variety of factors, including the components used, the production method, the packaging, and the storage conditions. This is stated on the product's label with a date stamp.

5.2 PREVENTATIVE ACTIONS

To manage the pathogen content of food, a system of Critical Control Points (CCPs) should be created. Methods such as predictive modeling and challenge testing can be used to track the progression of a disease. Product shelf life and food safety, on the other hand, are intricately intertwined. Food must be safe to eat for the duration of its shelf life. Asset quality control by preserving the food's visual and sensory characteristics. Examine the label to check whether any nutritional promises have been satisfied (Turner et al., 2020).

The Australia New Zealand Food Standards Code, which governs the ingredients and how they must be labeled, applies to all food sold in New Zealand. Products having a shelf life of fewer than two years must be labeled with a date. Food must be safe up to and including the specified date, according to the Code. Alternatively, you may choose one of the following options: There is a "use by" date provided. This method is utilized for perishable commodities that represent a health risk if consumed after their expiration date. It is illegal to sell or consume food that has passed its "Use by" date. It bears a "use by" date. This section contains items that did not appear on the preceding list. It is not against the law to sell food that has passed its "Best before" date.

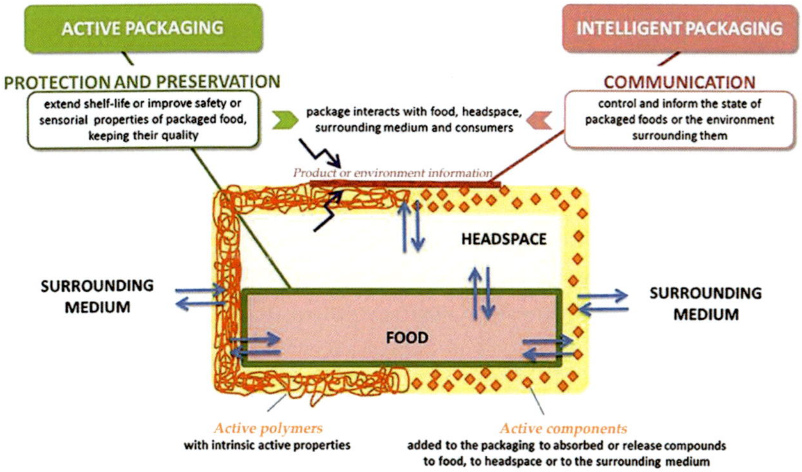

Figure 5.1: Good labeling on a product. (Source: Salgado PR, Di Giorgio L, et al., 2021)

The "Baked on" and "Baked for" date stamps are ideal for bread having a shelf life of fewer than seven days. The bread must be "Baked for" within 12 hours of being baked. What format does the data take? "Baked on," "Best before," and "Use by" all require a date or a reference to where that date is located on the box. The date must include the following information: The day and month must be specified if the product has a shelf life of up to three months. If possible, before January 24. Products with a shelf life of three months or more should contain the month and year at the very least. ON THE 5TH OF JANUARY, 2018. Except for the month, which can be written in letters, dates must be given numerically and chronologically. These must first be decrypted. In addition to the above-mentioned dates, "packed on" dates and packer's codes may be used. According to a Code requirement, item labels must contain comprehensive storage instructions where suitable to guarantee the product retains its freshness for a certain period. Storage needs must be met by distribution and retail networks, as well as individual houses. The food must be stored in line with the vendor's recommendations (Timakova et al., 2020). Once the product has been opened, it is critical to provide usage and storage instructions. When food is stored in modified atmospheres or under a vacuum, the sealed container has a significant influence on the shelf life of the product. In contrast to vacuum-packed sliced ham's one-month shelf life, which should be used within three to four days of opening when refrigerated.

Anyone who packages and sells date-marked food has a legal obligation to guarantee that their products do not lose quality over time. To guarantee that information about shelf life and storage requirements stays correct, all items should be labeled with this information. Repackers, secondary processors, food merchants, and supermarkets may also share part of the blame.

Every step in the food manufacturing process influences the quality and safety of the finished product. It is impossible to ensure food safety in the absence of a food management system capable of detecting and controlling hazards across the food chain. It is critical to realize the interconnectedness of all food chain nodes.

Primary Producers and Agronomists

Please make the necessary supplies available. Because of the varied quality of the raw materials, it may be difficult to develop a consistent end product. Raw material requirements may be set to ensure the quality of the end product.

Suppliers Are Also Considered.

Suppliers of cooling materials, such as ice, must, for example, ensure that their goods are secure and consistent. Additional guidance on alternate materials and how they affect shelf-life may be provided.

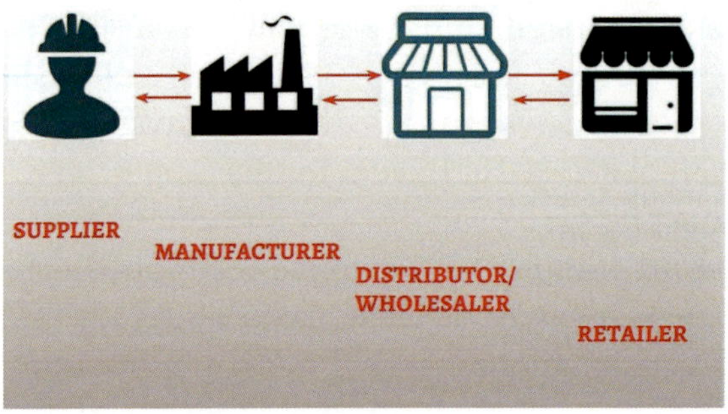

Figure 5.2: Difference between wholesalers retailers and distributors.

Manufacturers are exclusively responsible for choosing an adequate shelf life for their products. To reach this conclusion, all stages of the manufacturing process should be thoroughly scrutinized.

Distributors

It is normal practice to transport food from the producer to the shop. Food must be stored safely and securely, delivered on time, and in its original packaging. Concerns of food contamination from other items transported in the truck must also be addressed (Amani & Gadde, 2015).

Retailers

Before they reach the buyer, the items must be instantly stored and maintained. They must adhere to the directions supplied by the product's manufacturer.

As a result, it is critical to offer accurate information about storage conditions to consumers.

5.3 GUIDE TO CALCULATING THE SHELF LIFE OF FOODS

The two most prevalent ways are as follows:

This is the most widely used phrase. To guarantee that the food is safe to consume, store it for longer than its stated shelf life and inspect it often for signs of degradation. You can't create every product using the same procedure. You'll learn about the measures to take and the decisions you'll need to make in the following sections (Sutton et al., 2016).

A method of identifying a person.

This method can be used to estimate shelf life by utilizing accelerated storage and/or predictive microbiological modeling.

Several decisions will need to be taken throughout the investigation, and it may be necessary to seek professional advice.

Determine what is causing the food to deteriorate or become useless.

A variety of factors determine how long a product will stay on the shelf. Using the categories below, make a list of all how the product's quality and/or safety might deteriorate. Determine which components contribute to food shelf life extension. Keep the whole process in mind, from the purchase of ingredients and packaging materials through the end client usage. Keep in mind that owing to higher temperatures, certain things decay faster in the summer than in the winter (Sewald & DeVries, 2003).

The next step is to establish which tests are necessary.

Testing is required to ensure the product's safety and quality. Several tests do not apply to all commodities. Although lactic acid bacteria may be found in raw meat, they cannot be found in fermented raw meat (salami). To validate the product's safety, testing for food-borne illness-causing microbes such as Listeria may be necessary.

If a lab test is required, ensure that the lab is IANZ-accredited for the tests in concern. If you engage a lab to do storage temperature stability testing on your behalf, you must examine their records. In general, the following four types of tests are available:

The initial stage is to analyze sensory data.

Sensory assessment takes into consideration a food's fragrance, look, flavor, and texture. It may be used to notice and record important changes over time to calculate the shelf life of food. Food should be inspected in the location where it will be stored and consumed. This should ideally be done by a professional panel using well-established evaluation methodologies. Make ensure that the item being tested is safe to eat before using a taste panel. If feasible, samples should be frozen before testing to avoid changes in the texture or other features of the meal. This can be used as a baseline for future testing. If the meal cannot be frozen, a newly prepared sample should be utilized instead.

The military's use of microbiology

These tests can be used to assess the nutritional value and safety of a product. Yeasts, molds, and bacteria may all be examined over time to see how much and what kind of spoilage they create. If a culinary item is not on the restaurant's menu, it is up to you to decide where it should be placed. Microorganisms that might cause food poisoning must be detected to assure food safety. Depending on the product, there may be several tests that must be performed (Sumner & Jenson, 2011). The microbiological safety profile of a food specifies the types and levels of microorganisms that are permissible and harmful in the product.

5.4 CHEMISTRY

The quality of a product might deteriorate over time, which can be detected through chemical analysis. pH, headspace gas analysis, free fatty acid analysis, and total volatile nitrogen analysis are all examples of instrumental chemical testing.

Aesthetic value These tests check the texture of the product, examine the packaging, conduct "travel tests," and identify the best, best, and average retail settings, in addition to determining the best, best, and average retail settings. A "travel test" can assist in detecting potential transportation and handling hazards. Through this procedure, distribution and storage networks must provide things in the order they were intended (Angourakis et al., 2015). The whole supply chain, as well as the ultimate result, must be properly examined. This may be accomplished by employing a temperature data logger, which can be configured to capture temperature readings at certain intervals. The experimental design should take into account projected temperatures for delivery to retail outlets, business control, consumer purchase and transfer, and consumer storage.

Consider your product's shelf life.

To account for product variability, the experiment should be repeated several times. When it comes to product, procedure, and packaging, it's critical that the final result perfectly fits the prototype. Maintain thorough records of everything you use and do to analyze your results.

In step four, check the product's shelf life.

Samples should be kept in the same conditions as regular production samples throughout the study. If this is not practicable, a steady temperature and relative humidity should be maintained. These must be documented and verified regularly.

Because storage circumstances might change and product misuse can occur, you must account for the "real world" after estimating the shelf life under ideal storage conditions. Choose an acceptable shelf life above an ideal one, and provide a safety buffer. Limiting the product's movement and creating storage conditions can help to lessen the danger of product misuse (Sarkar et al., 2022).

Sixth Step:

Checking the Shelf Life of Your Products Acidity, taste loss, the presence of degrading organisms, and other properties discovered to be significant in shelf-life studies should be assessed on samples. Samples can be taken at many locations throughout the distribution and retail chain. If required, the original shelf life should be expanded to ensure safety. Furthermore, if the manufacturing or processing environment changes, the shelf life must be re-evaluated. The restricted components of this study can be identified and adjusted, resulting in longer shelf life. Investigating consumer complaints about things that do not meet expectations may reveal a reoccurring issue and the need to review the shelf life. Data collected during your shelf life research may be used to examine consumer complaints, manage manufacturing and distribution issues, and predict the product's shelf life. All test results must be documented and stored in a safe yet immediately accessible area.

5.5 FACTORS INFLUENCING THE SHELF LIFE OF A PRODUCT

It's time to determine how long I should keep my product on the market. There is no straightforward answer to this situation. The rate at which food decomposes varies greatly depending on the type of food. A variety of factors can contribute to quality deterioration.

5.5.1 An Increase in Microbial Communities

The growth of a wide variety of bacteria, yeasts, and mold in food can cause food spoilage or poisoning. Only foods that have already been contaminated during the manufacturing process, as well as any subsequent contamination that occurs during shipping, storage, or another handling, will be contaminated promptly by microorganisms. Temperature and storage time, in addition to the type of food, are important considerations (Calligaris et al., 2016). In meals, moisture is more likely to deteriorate than dryness. The presence of microorganisms or the adherence to established guidelines is frequently used to determine the end of shelf life. Sensory or biochemical degradation may serve as a warning sign that something has reached the end of its useful life.

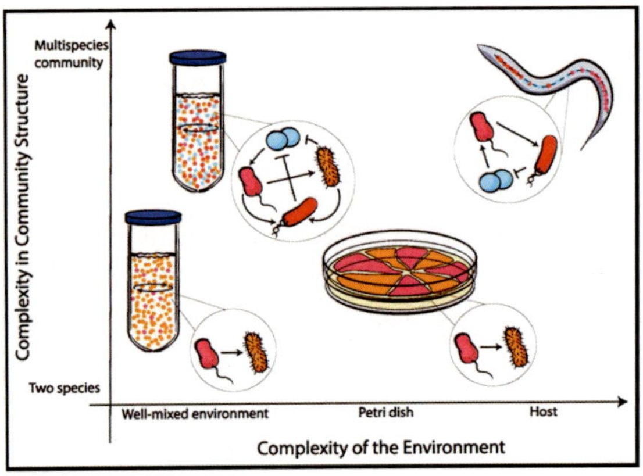

Figure 5.3: Microbial community.

5.6 BREAD

Sourdough fermentation affects all aspects of bread quality, including volume, texture, microbiological shelf life and staling, flavor, and aroma. A longer fermentation period allows cereal amylases, proteases, phytases, and pentosanases to work more effectively, resulting in increased acid production by lactic acid bacteria.

Figure 5.4: Bread.

As a result, sourdough bread is still popular in countries where rye bread is the most popular type of bread. The amount of dough acidification in rye bread is determined by two factors. Rye flour relies heavily on pentosans to keep the dough moist and retain gas due to its lack of gluten proteins. The partial hydrolysis of pentosanases during the fermentation process of the sourdough starter improves bread volume and texture while also increasing their solubility and water binding. Because of the polymeric gluten proteins in the dough, wheat sourdough makes water-insoluble pentosans soluble. Second, rye flour contains more amylase than wheat flour and has a lower gelatinization temperature. This means that amylase and gelatinized starch can coexist in a limited temperature range. Unless the pH of the flour is reduced, the amylase activity of rye flour in baking can compromise or even destroy the crumb structure (Castrica et al., 2021).

The low pH stimulates cereal proteases and phytases. Minerals in insoluble complexes with phytate are abundant in wheat and rye flours. The activity of cereal phytase enzymes is increased by acidification, which aids in the breakdown of insoluble phytates in bread dough. Through the hydrolysis of phytate, sourdough fermentation reduces mineral binding and increases bioavailability. This is unlikely to be required for a customer who consumes a nutritious, well-balanced diet. The American diet, on the other hand, may be deficient in trace elements, necessitating phytate reduction. There is no evidence that lactic acid bacteria phytases aid in phytate degradation during sourdough fermentation.

Wheat and rye flours are most active at pH levels ranging from 3.5 to 4.5. As a result, sourdough fermentation's high protease activity allows for a 5- to 10-fold increase in amino acid content. Amino acids are required for lactic and yeast metabolism, as well as heat reactions during baking. Excessive proteolysis, on the other hand, reduces the volume of wheat sourdough bread by damaging the gluten network.

Microbial metabolites such as acetic acid, ornithine, glutamate, and exopolysaccharides have a substantial influence on bread quality. Acetic acid is created when heterofermentative lactic acid bacteria employ fructose as an electron acceptor. Acetic acid, a volatile molecule, affects the aroma and flavor of the bread. Because acetic acid, rather than lactic acid, has antibacterial qualities, sourdough bread may be kept fresher for longer. When acetic acid is present at sufficient levels to prevent fungal bread rot, most people dislike its flavor of it.

These and other sourdough bacteria convert arginine to ornithine, which is required for fermentation. Because of the interaction between ornithine and 2-acetyl-1-pyrroline during baking, wheat crusts have a distinct odor. Experiments to boost sourdough's ornithine concentration boosted the superb, roasty crust scent even more. Wheat and rye proteins have the highest glutamine content, with gliadins holding up to 50% glutamine. Lactobacilli in sourdough convert glutamine to glutamate, giving sourdough its umami flavor. It was discovered that adding glutamate at amounts that matched microbial glutamate accumulation enhanced the sensory characteristics of the bread (Roccato et al., 2017). Even though all lactobacilli can convert glutamine, the quantity converted by sourdough lactobacilli strains varies. When yeast metabolism converts amino acids, flavor volatiles is generated; for example, leucine or isoleucine and phenylalanine are changed into methylbutanal and phenylethanol, respectively.

Lactobacilli in sourdough create exopolysaccharides in response to glucan sucrase and fructansucrase activity. These enzymes, which are found outside of cells or in cell walls, convert sucrose into polymeric fructans or glucans. Lactobacilli in sourdough generate fructans such as inulin, levan, and dextran, as well as reuteran and mutan. Every sourdough has at least one strain capable of producing exopolysaccharides. Exopolysaccharides are produced by sourdough fermentation in response to the amount of sucrose in the dough and other process factors. Exopolysaccharide dough yields more than 20g kg1 in experimental and industrial fermentation procedures. This quantity can offer more volume and texture to sourdough bread while also slowing the aging process. The production of dextran by *Leuconostoc* spp. during sourdough fermentation has been connected to the prolonged shelf life of Panettone. During the conversion of sucrose by glucansucrases and fructansucrases, heterofermentative lactobacilli release fructose and create extra acetic acid. Only a few heterofermentative lactic acid bacteria do not require fructose as an electron acceptor to improve acetate synthesis, and Weissella species that make dextran are excellent for increasing bread volume and texture with yeast.

When the bread is fermented in the presence of sourdough, bacilli and mold cannot form ropes and destroy the bread. The acidity in most sourdough bread is modest enough to prevent ropy decomposition. Experimental sourdough fermentations were demonstrated to limit the formation of fungal spores on bread. Each strain's antifungal capabilities differ. Because lactobacilli do not create inhibitory amounts of recognized antifungal metabolites during sourdough fermentation, active antifungal metabolites have yet to be found (Pratiwi et al., 2015). To some part, sourdough's antifungal capabilities stem from the substrate's generation of antimicrobial compounds and metabolites.

5.7 INTRINSIC FACTORS

5.7.1 Moisture Content

Microorganisms in food require a steady supply of fresh water to grow. Controlling the moisture content is one of the most well-known strategies for food preservation. Water activity is a term often used by food microbiologists to describe the water requirements of bacteria in foods or surroundings.

Fresh meat, vegetables, and fruits, like most other fresh foods, have aw values that are close to or at the optimal level of development for the great majority of bacteria. To manage aw in meals, solutes like salt or sugar, drying or baking, or the binding of water to macromolecular components in the meal can all be used. Polyhydric ethanol, amino acids, and other low-molecular-weight chemicals such as cellulose, protein, and starch outperform high-molecular-weight substances such as sugars and polyhydric alcohols (Peleg et al., 2017).

A wide range of variables determines how microbes react to aw. Bacterial growth and metabolite production may have a direct impact on the aw. Microorganisms require varied levels of aw depending on their environment to grow. Taxonomic categorization is one indication of microbial response. Many bacterial infections are treated at water activity levels more than 0.86 because, at an aw of 0.90 or higher, only one *S. aureus* can thrive and produce toxins. Keep in mind that these are only estimates since different solutes have variable powers to inhibit microorganisms at the same Aw value. When NaCl was the solute, for example, the lower aw limit for *Clostridium botulinum* type was used. When glycerol was used, the lowest aw limit was 0.92, and the growth was 0.94. When designing foods that use aw as the primary pathogen control mechanism, microbiological challenge testing can be performed to evaluate the efficiency of lowered aw concentrations that are close to the organism's growth limit. Various moisture monitoring methods may be more accurate depending on the kind of solute or humectant. Antibotulinum characteristics of pasteurized processed cheese spreads are known to be influenced by variables other than aw. Aw may be used in conjunction with other variables to further limit the spread of infection. Aw, measurements may not accurately depict the value in a microenvironment or at the interface of components in a multi-component meal. Aw should be characterized by both the food interaction and any possible microenvironment.

Where,

W1- Wight of the product before drying

W2- Weight of the product after drying

The quantity of moisture in a food product may impact many elements of its look, texture, and flavor, as well as its shelf life, freshness, quality, and resistance to bacterial infection.

Using less water in a food product can help to prevent microbial development and improve the shelf life of the product. As a result, the specified expiration dates of the objects may be considerably affected.

All of these elements have an impact on a food producer's success. Customers' satisfaction, loyalty, and, eventually, revenue and profitability are all dependent on a product's ability to retain specific attributes.

The impacts of each factor that can be impacted by moisture content on the food business are examined in depth in the sections that follow (Omotilewa et al., 2018).

The shelf life of a product is essential since it tells customers when they can eat it. When a product is damp, it is more susceptible to germs, which can lead to food deterioration and damage. Producers can now predict a product's shelf life with confidence, ensuring that customers are not exposed to spoiled food.

5.7.2 pH and Acidity

Fermentation or the addition of weak acids has been employed as a means of preservation since the dawn of time. Meat, fish, and vegetables are all naturally low in acidity, but fruits are high in acidity. For example, the white of an egg is alkaline.

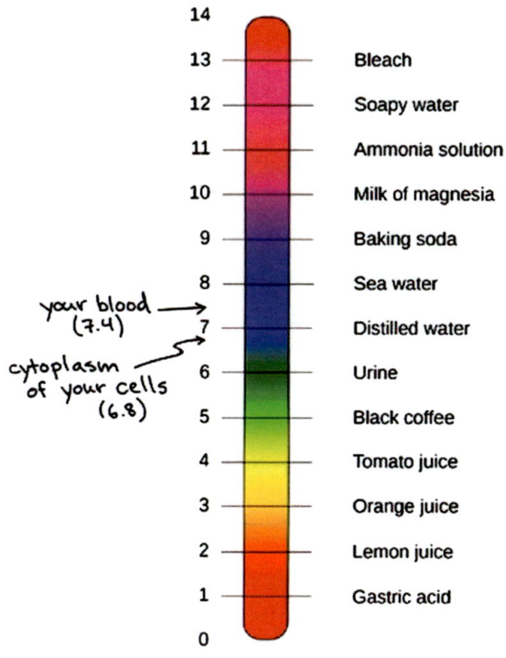

Figure 5.5: pH Scale: Acids, bases, pH, and buffers.

When it comes to meal pH, another crucial element to understand is pKa. The dissociation constant of acid is denoted by the symbol pKa. The pH at which the concentrations of dissociated and undissociated acids are equal in equilibrium is denoted as pKa. Strong acids are almost entirely dissociated in an aqueous solution when their pKa goes below 7. A hydrogen chloride (HCl) solution, for example, has a pH of 1.08, but acetic acid (HCO3) has a pH of 2.06. This feature is critical when employing acidity as a technique of food preservation (O'Brien & Bentley, 2015). Organic acids that have not been dissociated are more efficient preservatives than dissociated organic acids. Organic acids' preservative qualities can be improved by decreasing the pH of a meal. The kind of organic acid utilized has a significant impact on microbiological preservation and safety. For bacterial development in foods, a wide range of optimal, minimum, and maximum pH levels has long been recognized as scientifically reliable. The combination of other dietary factors, like pH, might stymie development.

Acidity, salt, temperature, and preservatives all interact with pH to prevent infections and disease-causing organisms from growing. The acidity level of the meal also affects toxic heat-treated food. This approach is recommended because germs may be destroyed with less heat when the pH is reduced. Another important aspect of a meal to consider when using acidity as a management technique is its buffering capability. The buffering capacity of a metal refers to its ability to withstand changes in pH. Bacterial development produces acidic or alkaline chemicals that quickly affect the pH of meals with little buffering capacity. Meat has a stronger buffering capacity than vegetables due to its vast diversity of proteins. Some foods' titratable acidity, such as salad dressings, is a stronger predictor of their microbiological stability than pH. The amount of standard alkali required to neutralize an acid solution is referred to as "titratable acidity." This titration determines the formation of hydrogen ions during the titration of undissociated acid. Titratable acidity is especially useful for meals that are either well buffered or extremely acidic. The presence of weak acids has no direct effect on the pH of a solution since they are more likely to stay undissociated.

In some circumstances, titratable acidity, rather than pH, can be used to assess the overall acid content of a meal. Illnesses are unlikely to originate or grow if the pH is less than 4.6; nevertheless, there are notable exceptions. Some illnesses can flourish in diets with pH levels lower than their growth minimums. Toxin generation has been reported by *C. botulinum*, although these experiments employed large quantities of inoculum, soy peptone, and the presence of *Bacillus* spp. These facts were judged irrelevant by the panel because they did not influence the meals reviewed in the study. pH fluctuations can also convert a diet into one that favors the development of illness. Several botulism epidemics have been connected to mold-infested foods, for example. The shelf life of a food item is greatly impacted by these elements. According to the panel, pH 4.6 is optimum for reducing spore-forming illnesses after performing a thorough literature assessment (Nicoli, 2012).

It is widely known that *Salmonella* spp. thrive in conditions with low pH, however, Chung and Goepfert observed that the limiting pH was considerably changed by adding an acidulant. Acids such as hydrochloric, citric, and acetic can change the pH necessary for salmonella development, ranging from 4.05 to 5.5. Atypically high quantities of Salmonellae inoculum were also identified in food systems. However, because the experiment was done up in a laboratory medium under ideal temperature and aw conditions and without competing bacteria, these scientists are wary about extrapolating their findings to food. When exposed to HCl at pH 3.8 and 30 °C for 1-3 days, six of the thirteen *Salmonella* spp. strains representing 12 serovars grew swiftly, but at pH 3.8 and 20 °C, they required 3-5 days to develop rapidly. Additional study reveals that acids with a pH of 4.5 can kill salmonellae, according to previous studies. As a consequence, the panel decided that the need for commodities subject to Food Code regulations to have a pH of 4.0 for *Salmonella* spp. was not scientifically justifiable. A thorough study of the scientific literature led the panel to determine that a pH minimum of 4.2 would be scientifically adequate to control for Salmonella spp. and other vegetative pathogens. As with other intrinsic qualities, the pH should be evaluated not only for each food component but also for each conceivable microenvironment when examining multicomponent meals.

5.7.3 Nutrient Content

There are a few nutrients that are required for microorganism growth and maintenance. Microorganisms require a wide variety of nutrients. Among these nutrients are water, which provides energy, nitrogen, vitamins, and minerals. These nutrients are present in varying amounts in a variety of foods. There are numerous health benefits associated with meat consumption, including the provision of essential nutrients such as protein, fats, minerals, and vitamins (Nirupama et al., 2010). There are numerous low-carbohydrate muscle meals available. Carbohydrates, proteins, minerals, and vitamins are all present in a variety of plant foods. Dairy products such as milk and cheese, as well as egg yolks, are high in nutrients.

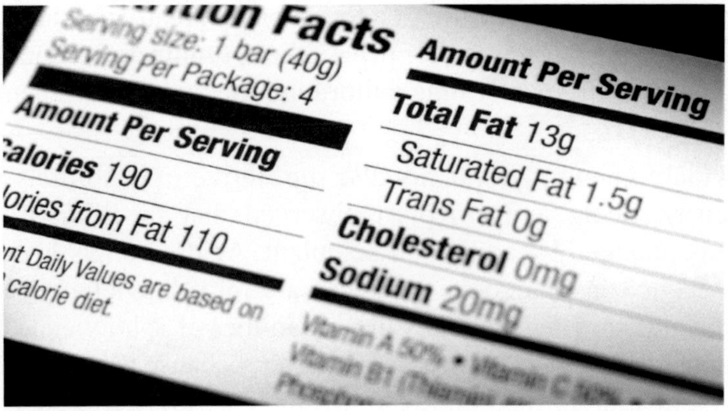

Figure 5.6: Nutrition Analysis.

Carbohydrates, ethanol, and amino acids are all efficient energy sources for foodborne microbes. Glucose is a simple carbohydrate that the vast majority of bacteria can metabolize. Certain animals are capable of digesting starch or cellulose derived from plant meals, as well as glycogen derived from muscle feeds. Certain microorganisms' lipids can be used to generate energy. The vast majority of microbes obtain nitrogen and energy from amino acids. Certain bacteria are capable of metabolizing proteases and more complex proteins. Methylamines and other nitrogen-containing compounds are also nitrogen-containing compounds. Numerous minerals are required for the growth of microorganisms. Minerals are found in a wide variety of foods due to their minimum requirement. Gram bacteria are unable to synthesize essential nutrients because they are more selective than other bacteria about what they require to survive and grow. Gram-positive foodborne pathogens such as *S. aureus* require a variety of nutrients to thrive (Nagi et al., 2012). Fruits and vegetables deficient in B vitamins are ineffective at supporting the growth of these bacteria. Gram bacteria can obtain all of their essential nutrients from a variety of different foods, including carbohydrates, proteins, lipids, minerals, and vitamins. *Salmonella enteritidis*, for example, requires a specific diet to thrive. Iron may inhibit *Salmonella enteritidis*. For example, antibacterial chemicals and a low concentration of free iron in the albumen of the egg prevent *Salmonella enteritidis* from growing to dangerous levels. Clay and Board discovered that supplementing an inoculum of *Salmonella enteritidis* in egg albumen with iron increased the pathogen's growth. Consuming nutrient-dense foods typically results in a higher concentration of microorganisms capable of effectively utilizing them. Simple carbohydrates and amino acids are typically consumed first, followed by their more complex counterparts. Due to the complexity of meals in general, multiple microorganisms can grow concurrently in a food. The availability of essential nutrients affects the rate of growth of an organism. Due to the abundance of nutrients in most meals, a wide variety of foodborne infections can thrive. That is why we cannot predict pathogen growth or toxin production reliably solely based on the nutritional composition of a meal.

5.7.4 Biological Structure

Foods originating from plants and animals, particularly when uncooked, include biological components that may block the entry and growth of dangerous microorganisms. Seed testa, fruit and vegetable skin, nutshells, animal hide, egg cuticle, shell, and membranes are all examples of physical barriers.

Pathogenic bacteria can adhere to the surface of plant and animal feeds or become trapped in the folds and crevices of the surface. Thus, undamaged biological structures can play a role in preventing bacteria from acquiring access to and multiplying. Numerous factors may affect the penetration of these barriers. Plant foods' maturity stage has an impact on the efficacy of protective barriers. Bacteria may enter through physical damage caused by harvesting, transportation, or storage, as well as insect infestation. During food preparation activities such as slicing, chopping, grinding, and shucking, physical boundaries are torn down (Maertens et al., 2012). As a consequence, depending on the essential qualities of the meal, the interior may get contaminated, resulting in growth. *Salmonella* spp. has been reported to thrive on the inside of cut cantaloupe, watermelon, honeydew melons, and tomatoes when given the appropriate time and temperature. Fruits reveal the capacity of hazardous microbes to break intact boundaries. While infections can linger on the external surface of fresh fruits and vegetables after harvest, they are seldom observed. Because foodborne bacteria lack the enzymes necessary to break down the protective outer barriers found on the majority of vegetables, development on undamaged surfaces is unlikely. This outer layer protects nutrients and moisture from entering. The formation of *E. coli* on the surface of watermelon and cantaloupe rinds is an exception.

When the protective epidermal barrier is disturbed, whether by physical trauma such as punctures or bruising, or pathogen breakdown, the survival of foodborne infections on produce rises dramatically. In these environments, germs multiply more rapidly, particularly at higher temperatures. Bartz and Showalter predicted and observed fruit infiltration using the general gas law, which stipulates that any change in the pressure of an ideal gas confined in a closed container of constant volume is directly proportional to the change in the gas's temperature. While the artwork by Bartz and Showalter displays a tomato, any fruit, such as an apple, may be viewed as a partially closed container. The internal gas pressure of the container or fruit falls as it cools, causing a partial vacuum within and allowing air to enter from the outside. Increased external pressure, such as that caused by immersing warm fruit in cold water, may result in pathogens infiltrating the fruit surface or cooling water. Bacterial internalization into fruits and vegetables can also emerge as a result of tissue ruptures or morphological anomalies inside the fruit, such as a calyx or stem scar (Manzocco, 2016). Despite the risk of infiltration, the panel found that there is insufficient epidemiological data to recommend freezing intact fruit. The egg is another example of a biological structure that shelters the perishable yolk from external microbial contamination when intact; nonetheless, contamination can occur by transovarial infection. Microorganisms on the egg's shell must be penetrated to infect its contents.

Additionally, egg whites include antibacterial qualities. Microorganisms can enter the egg via breaks in the inner membrane. Internalization is regulated by a myriad of characteristics, including storage temperature, relative humidity, egg age, and degree of surface contamination. For example, extreme humidity, moist and unclean shells, and a drop in storage temperature all enhance the danger of germs entering. The wash water should be 12 degrees Celsius warmer than the eggs to minimize germ penetration. Following that, rinse, dry, and refrigerate the eggs. The Food and Drug Administration has announced a final regulation demanding that all live Salmonella be removed before being provided to customers in raw shell eggs. To avoid *Salmonella enteritidis* generation, eggs must be kept dry and refrigerated below 7.2 degrees Celsius. As with other kinds of processing, food heating damages protective biological structures and affects pH and aw. Microbial infections may be able to proliferate as a result of these alterations.

5.7.5 Redox Potential

The oxidation-reduction potential of a substance can be calculated by multiplying its oxidation potential by its reduction potential (also known as its redox potential). In its most basic form, an electron can be gained or lost, and this is known as redox potential. Oxidation potential Redox potential is measured in millivolts. At pH 7.0 and 30 °C, the Eh of standard oxygen electrodes is +810 mV and -420 mV, respectively. Eh is affected by the pH of the substrate, which is typically measured at pH 7.0. Aerobes, anaerobes, facultative aerobes, and microaerophiles are the four basic groups of microorganisms that all require Eh. Pathogens found in food include *Aeromonas hydrophila*, *Clostridium botulinum*, *Escherichia coli*, and *Campylobacter jejuni*, to name a few. Anaerobes grow at voltages ranging from 100 to less than 250 millivolts, whereas aerobes grow at voltages ranging from 500 to 300 millivolts (Manzocco et al., 2010). Growth of the strictly anaerobic *C. botulinum* is slowed by Eh values above +60 mV, but growth is still possible at lower Eh values. The effects of salt and other dietary components on the Eh-development relationship can be significant. Toxins were produced in contaminated products that were stored at 15 °C for three days with an Eh of +200 to +250 mV. The primary oxidant of *Clostridium botulinum* is trimethylamine oxide, which also serves as an electron acceptor. The limiting Eh for the anaerobe *Clostridium perfringens* is close to +200 mV, but it rises as salt concentrations increase. Another factor to consider is the dish's poaching capability. A meal's Eh has "poising capacity," which is similar to buffering capacity if it can withstand changes in Eh caused by external influences. Active respiratory enzyme systems, as well as oxidizing and reducing agents, can all have an impact on a food's poisoning potential. Fresh fruits and vegetables have lower Eh values because they are still respiring.

No matter what kind of diet you follow, you can calculate the redox potential of the food. The redox potential of interface regions and microenvironments must also be considered when evaluating multicomponent meals. It is difficult to obtain precise measurements that account for the internal variations of the food as well as its equilibrium state. Morris recommends that the electrode be prepared and calibrated in such a way that the readings are stable and reproducible and that the food being tested be tested in such a way that the potential being measured is not affected. It is a mistake to rely solely on redox potential information to predict food safety or prevent the growth of certain microbes (Martins et al., 2008). The likelihood of a pathogen developing could theoretically be estimated using redox measurements in conjunction with other criteria. If you don't have access to additional resources, it can be difficult to obtain the most accurate results.

5.7.6 Water Activity

The capacity to estimate and regulate food shelf life is strongly dependent on water activity. The shelf life of a product refers to how long it will be safe, have the necessary sensory, chemical, physical, and microbiological attributes, and comply with nutritional labeling. Water activity, pH, redox potential, oxygen, preservative use, and processing/storage conditions all have an impact on shelf life. By measuring and controlling the water activity of foods, it is possible to predict which microorganisms will be potential sources of spoilage and infection, maintain the chemical stability of foods, minimize nonenzymatic browning reactions and spontaneous autocatalytic lipid oxidation reactions, control the activity of enzymes, extend the shelf life of nutrients and vitamins, and optimize the physical properties of foods. Every bacteria has a water activity threshold beyond which it cannot live. Controlling water activity minimizes microbial growth and allows some objects to be securely preserved without refrigeration (Lindström et al., 2006). Water activity boosts deteriorative chemical reaction speeds because it functions as a solvent, may be a reactant, and can affect the mobility of reactants through viscosity. Any of these variables, alone or in combination, can result in quick degradation and a reduction in product shelf life. As aw grows, non-enzymatic browning processes accelerate, peaking at 0.6 to 0.7aw. Lipid oxidation is at its lowest at about 0.2 to 0.3aw.

Water activity is a ratio of the vapor pressure of water in a material (p) to the vapor pressure of pure water (p_o) at the same temperature.

Moisture migration is a primary cause of texture loss and degradation in food. Temperature changes in transportation and storage affect microbiological, chemical, and physical qualities since water activity is a function of temperature. It's vital to establish what a product's critical water activity levels are, as well as how sensitive it is to changes in water activity. According to moisture sorption isotherms, the highest chemical stability is observed at the monolayer moisture content. Changes in water activity, such as water absorption when a product is exposed to a high humidity environment or water loss when a product is exposed to a low humidity environment, induce undesired product changes and lower shelf life. Physical changes include the loss of crispness in dry items, powder caking and clumping, and wet products losing water and becoming rough and chewy (Liu et al., 2018). Acceleration of chemical breakdown or the chance for microbial growth is two more negative outcomes. Understanding and maintaining a product's critical water activity level through suitable packaging can help to extend its shelf life. If you retain dry powders without a moisture barrier or anticaking agent, they will undoubtedly be cake. Most breakfast cereals are less prone to moisture absorption and don't need high barrier levels. Moisture loss is necessary for some baked dishes, but if there is too much moisture in the container, microbial growth could ensue. If the water activity of the individual components cannot be shifted closer together, multi-component items may require specific packaging or barriers to limit moisture transfer. It is achievable to calculate shelf life based on the suitable barrier qualities of the packaging material or dictate the barrier specifications of the package to attain the desired shelf life by understanding the critical water activity limitations of a product (Kanatt, 2020).

5.8 NATURALLY OCCURRING AND ADDED ANTIMICROBIALS

Some foods include antibacterial chemicals that assist to make them microbiologically stable. Some foods may contain essential oils, tannins, glycosides, and resins, as well as other plant-based antimicrobial components. Cinnamon includes cinnamic aldehyde as well as eugenol, whereas sage and oregano both contain thymol and carvacrol in addition to eugenol. Eugenol may also be found in cloves and allicin in garlic. Plants include phytoalexins and lectins, which have antibacterial characteristics. By linking them to the protein, lectins can be linked to numerous polysaccharides, including glycoproteins on the cell surface. The binding of lectins with bacteria may result in antibacterial effects. Antibacterial activity is diminished in prepared meals due to the low concentrations of these compounds. However, combining these compounds with additional formulation components may increase formulation stability. Antimicrobial components may be present in some animal-based diets. Lysozyme is a naturally occurring component in fresh meat, poultry, and seafood, in addition to lactoferrin, conglutinin, and the lactoperoxidase system (Kuijt, 2009). Lysozyme is a protein that can break down bacterial cell walls. There are three distinct antibacterial components in the lactoperoxidase system of bovine milk: lactoperoxidase, thiocyanate, and hydrogen peroxide. Psychotropics that generate lactoperoxidase, such as pseudomonads, are particularly vulnerable to this mechanism.

As a consequence, an improved version of this strategy has been proposed to improve the quality of raw milk maintained in developing nations with limited refrigeration access. Antimicrobial compounds generated from animals, like those derived from plants, do not influence the ambient shelf life of food. Certain food processing processes have also been proven to produce antibacterial compounds in food. Smoking fish and meat produce antibacterial chemicals on their surfaces. As they heat, sugars condense with amino acids or peptides, which may have antimicrobial effects. Smoke condensate phenol is an antibiotic that kills bacteria while also lowering the pH of the surrounding surface. Some processors employ liquid smoke to decrease the pH of the unsliced shelf-stable food's surface. Several fermentation processes can produce bacteriocins, antibiotics, and other antimicrobial chemicals. Certain bacteria create proteins or peptides that inhibit the development of pathogens prevalent in the environment. Lactic acid bacteria are the most prevalent producers of bacteriocins. Nisin, a lantibiotic produced by numerous strains of *Lactococcus lactis*, is one of the most investigated bacteriocins. More than 50 nations have approved the use of nisin in cuisine. Nisin was first used in the kitchen to protect Swiss cheese from *Clostridium butyricum*. The polypeptide nisin poisons most gram-positive bacteria and fungus, however other organisms and fungi are unaffected. Nisin can be added to food as a preservative or used to enhance it with starting cultures. There is evidence that Nisin when interacting with heat, can limit the growth of spore-forming bacteria, and it has been successfully utilized in processed cheese recipes to accomplish this purpose. When nisin is added to a Fo technique for low acid canned foods, for example, a Fo of 3 can be attained for thermophilic spore inactivation (Chang et al., 2022). Several bacteriocins and natural antimicrobials have been found, but their practical application as food preservatives has been limited due to their narrow spectrum of activity, incompatibility with food formulas, or regulatory status.

To improve the shelf life of food and/or prevent germs, a wide range of chemical preservatives and additives can be used, either alone or in combination. These preservatives may be restricted by food regulations in a specific country or region. When choosing a preservative for a given culinary application, a variety of criteria should be considered. It is excellent to have antimicrobial activity against degrading organisms and diseases that are likely to be found in food. The preservative must be active both during storage and during the anticipated formulation conditions. This addition should not affect normal microbial activity in food, such as cheese ripening or leavening in baked products. Antimicrobial agent interactions or synergistic effects may have an impact on formulations. A pH relationship is one example of a possible connection. The pH of various preservatives is an essential element in their efficacy. There are several aspects to consider, including the kind of food, preservatives, enzymes, processing temperature, storage conditions, and partition coefficients. Food preservation can be accomplished by the use of preservatives in combination with other physical and chemical components of the food's composition. Leistner purposely designed the hurdle concept to discover these results. It is predicated on the notion that a combination of inhibitory chemicals, even if ineffective on their own, can be useful when combined. The hurdle concept is commonly used to characterize the anti-botulinum stability of different shelf-stable processed cheese compositions. Even though individual needs do not support this approach, these things may be preserved at room temperature for lengthy periods (Dawson et al., 2018). Other considerations, such as the possibility of antimicrobial migration to other meal components and the diverse nutritional content at the interface, may influence antibiotic efficacy in combination meals. Many food compositions do not require refrigeration for microbiological stability or safety due to the use of preservatives or the hurdle concept. In the lack of an established microbiological model, the microbiological safety of these items may be difficult to determine. Microbiological challenge tests are the most effective technique to establish if these formulations are suitable for non-refrigerated storage in the great majority of these circumstances.

5.9 COMPETITIVE MICROFLORA

Pathogen development in temperature-sensitive foods is influenced by several factors, including internal and external characteristics. A range of biological factors of individual animals impacts the diversity of microbial species in food. Two such examples are the growth rates of distinct microbial strains and reciprocal interactions or impacts among species in mixed populations.

5.9.1 Growth

In a feeding environment, growth and development may be expected. Genetic variables impact lag phase, generation period, and cell output. If metabolic waste products accumulate, a species' growth may be hindered. If species that can utilize the restricted metabolic product as a substrate create an alliance or succession. Because of the complex ongoing interactions between ambient components and bacteria known as association, each meal has a unique flora at any given time. Changes in the microbial population occur throughout time, which is referred to as succession. Two notable instances of this phenomenon are food rotting and microbiological degradation. When metabolically active species are present, changing flora dominance is unavoidable. These interactions can be antagonistic or synergistic depending on whether they promote or inhibit growth.

5.9.2 Competition

In food systems, antagonistic processes are often engaged when fighting for nutrients, attachment/adhesion sites, or adverse environmental changes. Because of their natural biota, *Staphylococcus aureus*, *E. coli*, and *Salmonella typhimurium* cannot infect frozen pot pies. A meat market is another site where raw ground beef is utilized. Despite the presence of *S. aureus* at low levels in this product, no staphylococcal enterotoxin is generated. The most likely culprit staphylococci are unable to compete with the Pseudomonas-Acinetobacter-Moraxella trio identified in this meal. If a species with a high metabolic rate preferentially depletes resources, it can hinder the growth of other organisms. The existence of necessary anaerobes in vacuum-packed, chilled fresh meats is influenced by carbon dioxide or a lack of oxygen. Staphylococcus germs, for example, carry very little nourishment. As a result, the presence of *Pseudomonas* spp. and Coliforms renders the staphylococcus inactive. Staphylococci are not the only Micrococcaceae genera that may take extra nutrients (Kendall & Dimond, 2007). Staphylococci are repressed when streptococci reduce their nicotinamide (niacin) and biotin supplies. *Staphylococcus aureus* has a difficult time competing in both fresh and frozen meals. Biota in the food chain employs antagonistic interactions, nutritional competition, and environmental change to change conditions so that *Staphylococcus aureus* cannot flourish and limit staphylococci growth. External and internal variables, like food, can influence the outcomes of competition.

5.10 THE INFLUENCE OF GROWTH IS LIMITED

Pathogens such as Enterobacteriaceae, yeast, and micrococci have evolved new growth-stimulatory pathways. Growth-promoting elements can have a significant impact on the development of a typical flora. There are several alternatives available, with the following being a few of the most common: Other species can access and use a single creature's metabolic activity. Microorganisms may grow in a variety of pH conditions. Lactic acid bacteria, for example, thrive in acidic environments such as those found in naturally fermented foods. *C. botulinum* has been reported to flourish on high-acid diets, which elevate the pH of the food. Changes in the Eh or aw content of the food ingested may have an impact on symbiosis. *C. perfringens* can be utilized to lower the redox potential of newly dead tissues, allowing for the growth of more important anaerobic microorganisms. Some links can only be fully appreciated when both species are present. Microorganisms in temperature-sensitive foods can be controlled using the barrier concept.

5.11 EXTRINSIC FACTORS

5.11.1 Types of Packaging/Atmospheres

The antibacterial characteristics of gases at ambient and sub-ambient pressures have received a great deal of attention in the scientific literature. Microbes are kept at bay by gases in two ways. They can be directly detrimental as a first stage, impeding development and reproduction. Certain microorganisms can be directly harmed by carbon dioxide, ozone, and oxygen (Howey, 2015). This inhibitory mechanism is influenced by chemical and physical properties, as well as the aqueous and lipid phases of the meal. If exposed in large amounts, anaerobic bacteria can be destroyed by the oxidizing radicals produced by O3 and O2. Carbon dioxide, when present in adequate quantities, is effective against obligatory aerobes as well as other bacteria. A second way of avoiding disease transmission is to modify the ecology of the microbial habitat through changes in gas composition. The competitive landscape shifts as the environment change.

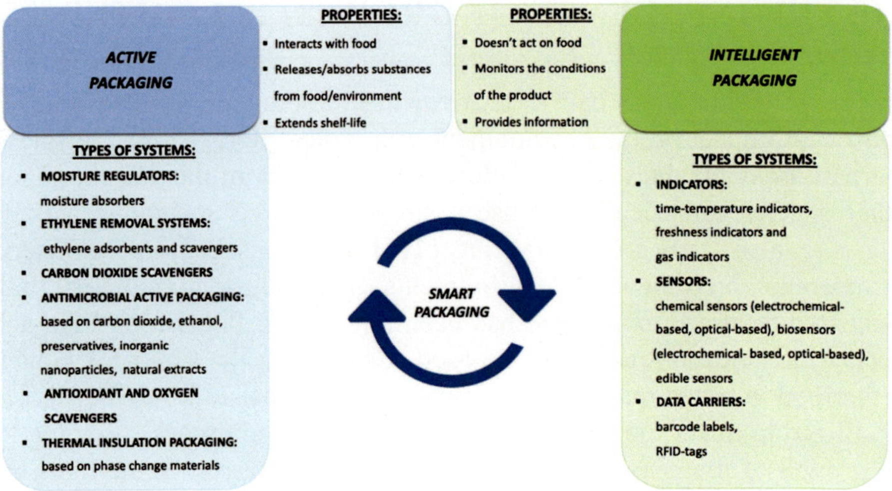

Figure 5.7: Types of packaging/atmospheres. (Source: Drago E, Campardelli R, et, al.,2020)

The presence of another organism in the environment may promote the proliferation of one germ. Depending on the pathogenic bacteria and the substrate, this impact might be beneficial or harmful. By substituting nitrogen for oxygen, an indirect antimicrobial effect may be proven. Traditional approaches for suppressing microbial growth are used, with the majority of these strategies relying on temperature. Modified atmosphere packing, controlled atmosphere packaging, controlled atmosphere storage, direct carbon dioxide addition, and hypobaric storage are all instances of technology. The use of packaging with a modified or controlled environment can greatly prolong the shelf life of some items. CO2, N2O, and Ethanol can all be used in a MAP application (De Corato, 2020). Because CO2 is more soluble at lower temperatures, its inhibitory effects grow more prominent as the temperature drops. When food is exposed to carbon dioxide, the pH of the food decreases. Nitrogen's lack of antibacterial characteristics is because it is an inert gas. This method, which eliminates oxygen from food packaging alone or in conjunction with CO2, indirectly inhibits aerobic microbes.

Antimicrobial conditions have been used to preserve fruits and vegetables, raw meat and poultry, dairy products, eggs, and a range of ready-to-eat meals. There are variously internal and external factors that must be considered for an antibacterial environment to be successful. Packaging, the microbiological quality and safety of the product, and the biochemical composition of the food all interact to influence how much microbiological quality and safety may be enhanced through various aspects. Temperature is the most critical component in antimicrobial atmosphere effectiveness since it directly impacts growth rate and indirectly through variations in gas solubility. Packaging design, particularly the volume ratio of product to headspace, influences microbial suppression at practical food storage temperatures. In terms of microbiological development, the barrier qualities of the package have a significant influence on how long selected modified atmosphere gases to remain in contact with the product and how rapidly oxygen enters the package in MAP. The overall inhibitory influence of antimicrobial gases is regulated by a range of parameters, including water activity, aqueous phase salt concentration, pH, and dietary fat content. The meal's physical and chemical qualities, as well as its temperature, all impact the inhibitory gas's solubility. As salt concentrations increase, CO2 solubility decreases. Bacterial deterioration is the major safety hazard related to the use of MAP or other technologies to enhance the shelf life of foods (Hedegaard & Skibsted, 2013). Organoleptic quality items may be harmful if there are no symptoms of decaying microorganisms. In foods exposed to a range of atmospheres, the effect of spoilage bacteria losing their competitive inhibition is particularly obvious. It is feasible to utilize antimicrobial settings and other techniques to improve food quality and safety.

5.11.2 Spoilage Causes Other than Microbial

There are various routes through which nutrients and quality can deteriorate. Even if the product is safe, these modifications may suggest that it is no longer suited for its intended usage. Nutritional loss, browning, and rancidity can come through the buildup and loss of moisture. As soon as dry foods come into touch with water, they become subject to microbial deterioration. Bad smells, color changes, browning, and nutritional inadequacy could arise from chemical modifications. Rancidity, vitamin depletion, and loss of natural color are all possible results of exposure to light. Different degradation mechanisms are enhanced or slowed down by temperature variations. A food's quality could decline if it is affected physically, such as by bruising or cutting. Food degradation can be induced by both microbial and non-microbial breakdown processes when packaging is destroyed. Grease may seep out of a pinhole in a container and germs can enter via a crack in a plastic bag, for example. The soil has also been contaminated by different sorts of rodents and insects. Smells and tastes emerge as a result of food being stored near other goods with strong odors or after a product has been tampered with.

5.11.3 Be Able to Spot the Item

Assessors must exhibit a comprehensive grasp of the product they are reviewing. Three primary types of food items exist:

These foods have a short shelf life because they are prone to microbiological and enzymatic decay. The shelf life of the product is determined by conducting daily measurements.

Natural preservatives may be employed in semi-perishable foods, which have a short to medium shelf life, or they may be stored without them.

To establish the shelf life, measurements are done every week. Foods with a lengthy shelf-life owing to heat treatment or storage are described as "extremely stable foods." Weekly or monthly statistics are utilized to determine shelf life.

5.11.4 Understanding What Affects Product Quality

Compositional and environmental factors both influence shelf life. As a rule of thumb, food composition includes factors such as water activity; pH; total acidity; acid type; redox potential; available oxygen; nutrients; natural microflora and microbiological counts; preservative use; natural biochemistry; preservatives; reactants; inhibitors; and catalyst concentration (Hosseini & Jafari, 2020).

Processing, storage, and distribution are all impacted by environmental characteristics such as the time-temperature profile, headspace pressures, temperature controls, relative humidity, light exposures, ambient microbial counts, and the atmospheric composition of the environment.

5.11.5 By Applying a Few Basic Principles, Find the Essential Quality

The client's standard may be higher than the government's. In government legislation, microbiological safety is the primary concern; Legislation from the government is characterized by this:

Consumers – are the finest evaluators of the critical quality of each food item. Competitor's product - the important quality may be connected to a competitor's product

You should be on the lookout for the Concerted Series of activities.

Understanding and identifying the mechanisms underlying the rotting or loss of beneficial features rely on these biochemical/physicochemical interactions. For example, food decomposition can be produced by microbial degradation, chemical or enzymatic activity, such as rancidity or browning, or by moisture or another vapor movement (Derkach et al., 2020).

CHAPTER 6

THE SHELF LIFE OF PACKAGED FOOD PRODUCT QUALITY

CONTENTS

6.1 INTRODUCTION

To the average customer, a food's shelf life refers to how long it may be stored before it expires, as shown on the "best-before" label. That is, food has a shelf life until it becomes unfit for human consumption. The shelf life of a food product may also be described as the amount of time it can be stored and shown while maintaining a certain level of quality or achieving a specific purpose. As a result, determining food safety solely on shelf life may be difficult. When a product's shelf life expires, it must meet a set of quality requirements to be safe for human consumption (El-Anany et al., 2009). Pasteurized milk that has been properly maintained and is free of microorganisms can be stored for several days after its expiration date. If you keep food over its expiration date, you increase your chances of getting food poisoning. This is especially true for bacteria-infested meals. Many items' shelf life and expiration dates are typically the same.

Packing materials may assist increase a product's shelf life, and this chapter describes how to do so. To increase product shelf life, the author investigates how packaging affects both product quality and shelf life. Packaging can shorten a product's shelf life. Toxic chemicals might have migrated into the meal, or food components could have migrated into the container, which could explain this. As a result, this chapter investigates the negative impacts of insufficient packing on product quality and shelf life. The process of achieving an agreement on a shelf life specification is rarely simple. For customers, merchants, distributors, producers, and farmers, for example, shelf life is the most critical concern. Customers want to know that they can rely on the things they purchase and that the quality is satisfactory. Customers are generally drawn to items with the longest remaining shelf life because they feel freshness is an essential aspect of product quality.

The most essential aspect of shelf-life testing is how individuals keep and consume their items. Retailers must ensure that their items meet or exceed their consumers' expectations in order to keep them coming back for more. Maintaining a product's shelf life should be updated so that it may be disseminated and sold while also giving the customer a suitable length of time to enjoy the product. Manufacturers must be able to demonstrate that the shelf life information they supply is correct when it comes to shelf life. To make matters worse, they are under intense pressure to supply items that fulfill the store shelf life criteria, which determine whether or not a product is stocked. Extending the life of products is a major driving force behind product and packaging innovation. Modified atmospheric packaging (MAP) and active packaging, for example, are two new forms of packaging that have been created to aid in the development of novel preservation strategies (Fan et al., 2009). It is hard to exaggerate the value of packaging in extending shelf life and maintaining freshness. The IFST rules define shelf life as the period that a food product stays safe, preserves the necessary sensory, chemical, and physical qualities, and microbiological characteristics, and adheres to any nutritional data stated on the label (1993).

This definition allows "desired qualities" to have a part in how long a product's shelf life can endure. As a result, this is critical since, after safety problems have been addressed, quality is a commercial issue that has a direct influence on marketing strategy. During the product development process, the shelf life of a product may be evaluated. Because packaging can be used to regulate shelf life-limiting activities or to limit the shelf life of the product itself, packaging requirements for the product must be addressed early in the product development process. Throughout the shelf life testing process, reprographic samples of a final product are stored in conditions that resemble those encountered during manufacture, distribution, and consumption. Once microbiological safety has been established, quality problems can be addressed. This can be determined in certain situations by microbiological counts, chemical requirements, or taste assessments (Galić et al., 2009). Most of the time, it will be a combination of these. Long-shelf-life commodities should be estimated using indirect or predictive approaches. The most typical approach for extending shelf life is to raise the temperature; however, other methods impacting product stability, such as humidity, shaking, or exposure to sunshine, can also be utilized. These studies are usually product-specific, needing a thorough understanding of the product and its behavior under accelerated environments. If this method is utilized, there is a possibility of chemical reactions or microbial development that would not be feasible under regular storage circumstances.

Predictive modeling, which employs mathematical models to estimate the product's shelf life or the amount to which it possesses a shelf life-limiting attribute, can also be used to accelerate shelf-life testing. When a meal's relative humidity is set to ERH, it will not absorb or lose moisture to the surroundings (value often expressed as Aw, the water activity). Mold-free shelf life and ERH have been discovered to be related to a variety of bread products. To anticipate the product's mold-free shelf life, use conversion factors to determine the number of sucrose equivalents provided by its constituents. When it comes to food microbiology prediction, mathematical models specify microbial growth under specified conditions such as temperature, pH, or preservative concentration. Two food microbiology models that may anticipate the future include the USDA Pathogen Modeling Program and the FORECAST Service of the Campden & Chorleywood Research Association (formerly Campden BRI). When microbial population increase is the limiting condition for shelf life, these models may be used to anticipate and quickly examine the influence of changes in product formulation on microbial growth and shelf life (GÁMbaro et al., 2006). Shelf life is not mandated by European Union law or regulation, and it is not regulated how shelf life is established.

According to Directive 92/59/EEC, producers are entirely responsible for ensuring that their goods are safe for customers. Pre-packaged goods must contain a minimum durability date (best before or best before the end) or, in the case of especially perishable commodities from a microbiological standpoint, a use-by date, according to the EU Food Labeling Directive (79/112/EEC and integrated into 2000/13/EC). Minimum durability dates are defined as "the period beyond which the product retains its distinguishing features when properly kept." The producer determines this date (and hence the shelf life). Because those in charge of food labeling are best equipped to judge the quality of the product, it is up to them to determine if a "use by" or "best by" date is necessary. Only authorized staff have the authority to update or erase the expiration date on food goods, and it is illegal to sell outdated food. If additives are necessary to meet the stated shelf life, additive regulations (which additives can be used and in what amounts) are critical. This is an unintended consequence of other pieces of law. Commodities with a restricted shelf life due to temperature-controlled procedures have a substantial influence on the maximum storage and transit temperatures for chilled and frozen items. Those that manufacture, import, distribute, or use food contact materials must ensure that their products are compliant with applicable EU regulations before selling them (EU). This rule does not apply, according to the Food and Drug Administration (FDA) in the United States, which is well-known to many businesses. In the EU, there is no precise definition of pre-approval or licensing of food-contact items (FSA, 2007). Food contact materials are controlled by legislation to guarantee that no contaminants from these materials are incorporated into foods and endanger human health or food quality. During the preparation, processing, packing, and transportation of foods, a variety of substances may come into contact with them.

The European Commission's Regulation No. 1935/2004 defines food contact materials (EC). Council Directive 89/109/EEC, a broad European Union rule, is replaced by EC Regulation 1935/2004. Food contact materials and other items that may come into contact with food must be manufactured and used in such a way that their components do not transfer to foodstuffs in proportions that could pose a health risk: foods and beverages that change their nutritional composition are harmful to human health. Alter the organoleptic properties of the food or drink EC Regulation 2023/2006, which is linked to Regulation 1935/2004, provides particular GMP criteria (EC, 2006). This rule, which specifies particular standards for methods involving the application of printing inks, also applies to the non-contact side of a material or object. Regulations influence migration into foods as well. They adhere to European Council and European Commission directives (EC, 2009). Directive 82/711/EEC defines the fundamental criteria for monitoring the movement of components of plastic materials and products designed to come into contact with foodstuffs in the EU, with further changes in Council Directives 93/8/EEC and 97/48/EC. Food simulants are established in European Parliament and Council Directive 85/572/EEC to be used in the study of migration from plastic food contact materials and articles. Plastic goods that come into touch with food are subject to a variety of laws. In 1990, the European Commission enacted Directive 90/128/EEC. More changes were made, all of which were incorporated into Directive 2002/72/EC. It has undergone various adjustments, according to European Commission Directive 2008/39/EC. Diverse instructions address a wide range of auxiliary packaging components, including regenerate-cellulose film, elastics/rubbers/paper/board/surface coatings such as paraffin/microcrystalline wax/ion exchange resin (Galovičová et al., 2021). The maximum amount of components permissible to migrate from plastic materials and articles into food is limited to a 10 mg/dm2 contact area as a general overall migration limitation.

The Polymers Directive 2002/72/EC and its five revisions set migration restrictions for certain monomers and starting materials to guarantee that only allowed monomers and starting materials may be used in food contact plastics. These constraints limit residual levels in the finished product. The many components that come into contact with food are governed by several EU directives and rules.

6.2 SOME OF THE ATTRIBUTES THAT MAY AFFECT PRODUCT QUALITY AND SHELF LIFE

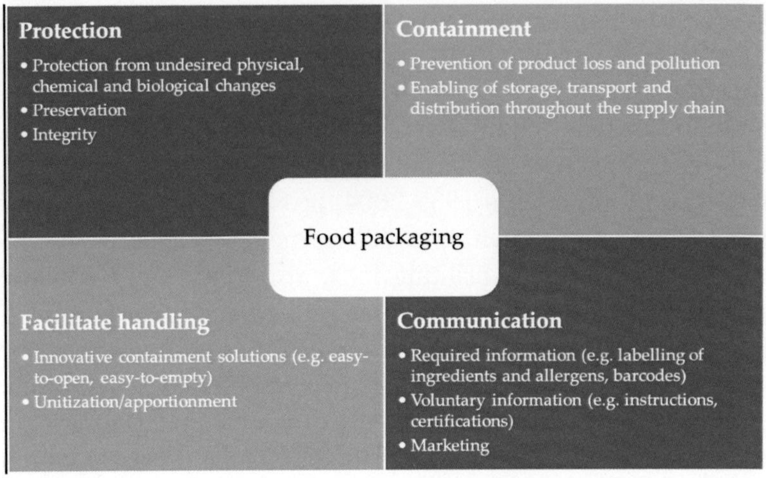

Figure 6.1: Factors affecting the shelf life of products. (Source: Wohner B, Pauer E, et, al., 2019)

Many foods have a short shelf life due to the presence of specific or "key" features that can be predicted during product development. This can be based on prior experience or observations with similar products, or it can be based on an examination of the product's intrinsic characteristics (intrinsic factors), the environment in which it will be used (extrinsic factors), and the shelf life limiting processes' that this combination of intrinsic and extrinsic factors is likely to produce. Intrinsic features are those that emerge spontaneously from the makeup of the final product and include the following: Water activity (water availability), pH/total acidity, acid natural microflora species, and microbial communities that remain in the end product are also factors to consider (Aw). The redox potential of oxygen is accessible (Eh) when other than salt, spices, and antioxidants are added to a product's native biochemistry/chemistry makeup. Interactions between packages, such as tin pickup and migration because subsequent processing seldom compensates for poor-quality raw materials, raw material selection is critical for managing inherent variability (He et al., 2018).

Extrinsic factors are those that emerge as a consequence of a product's interactions with its environment during its existence. The following are a few examples: Temperature control during storage and distribution–temperature profiles throughout all processes during storage and distribution, and the relative humidity (RH) is measured. Light (UV and IR) exposure during storage and distribution customer support packaging the composition of the gaseous environment Quality and Shelf Life of Packaged Goods Product packaging may have an impact on several extrinsic factors, and numerous innovations in packing materials have been motivated by the goal to reduce the impact of these external influences while extending shelf life. In certain circumstances, simply having the correct packaging can help enhance shelf life, such as by constructing a full light and oxygen barrier. In the great majority of cases, however, it is simply one of the numerous impediments that, when combined, diminish shelf life. The interaction of internal and external components impacts the likelihood of shelf-life-degrading reactions or processes. Shelf life-limiting reactions or processes are classified into three types: chemical/biochemical, microbiological, and physical. While the outcomes of these processes are rarely mutually exclusive, these categories provide a useful starting point for discussion. The consequences of these components are not always negative; in some cases, they are required for the creation of a product's desired characteristics.

6.3 EXTRINSIC ATTRIBUTES AFFECTING THE SHELF LIFE OF PACKAGED FOODS

Product Characteristics

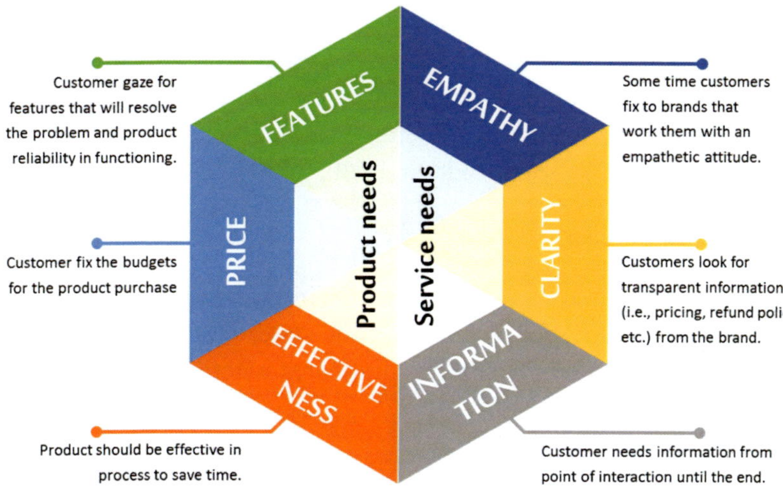

Figure 6.2: Product characteristics. (Source: Malik H, Afthanorhan A, *et,al.*, 2021)

Intrinsic elements include product attributes such as formulation and processing conditions. Intrinsic factors are qualities of a final product that occur as a result of its composition. They include the following:

- Total acidity/pH
- Counts of microorganisms and natural microbial flora in the end product
- The availability of oxygen
- Suggestions for cost-cutting initiatives (Eh)
- The inherent biochemistry/chemistry of the product
- Addition of preservatives (e.g. salt, spices, antioxidants) (e.g. salt, spices, antioxidants) (e.g. salt, spices, antioxidants)
- Product creation

6.3.1 Environmental Factors

An external element is an environment in which the product is carried and kept. To have extrinsic impacts, a product must come into touch with its environment at some point throughout its life cycle. Among the services they provide are the following:

6.3.2 Temperature

The temperature of the food and how it is packaged are two elements that impact the rate of degradation. The principal sources of cooling in refrigerated display cabinets are conduction and convection. This concept makes use of fluorescent lights, which give illumination while also producing heat. Aluminum foil offers substantial advantages in these contexts due to its high reflectivity and conductivity (Goldman et al., 2002).

6.3.3 Humidity

When discussing the amount of water vapor in the air, people use the term humidity. When water vapor is in a gaseous state, it is usually invisible to the naked eye. The degree of humidity indicates whether there is rain, dew, or fog in the air.

The humidity of a system is affected by how hot or cold it is. If there is the same quantity of water vapor in the air, cold air is more humid than hot air. The dew point is a comparable metric. As the temperature rises, so does the amount of water vapor necessary to attain saturation. When the temperature of a parcel of air falls below a specific point, it approaches saturation without adding or subtracting water mass. The quantity of water vapor in a particular volume of air can vary greatly. The relative humidity of the surrounding environment can influence the water activity of the food unless the container offers an efficient barrier to water vapor. Several types of flexible plastic packaging materials are moisture-resistant, however, none are impermeable.

6.3.4 Atmospheric Gases

As a result of the gases present in the air surrounding the food, the atmosphere within the packaging is continually changing. The vacuum packing, which removes air (and hence oxygen) from a box before sealing it, can prevent aerobic microbes from multiplying, which is one of the most fundamental means of altering the environment. A gas, such as CO2 or N2, is pumped into the box before it is closed to create a "modified atmosphere" (MAP). There are numerous methods for extending product shelf life, such as employing increased CO2 concentrations to prevent microbial development. MAP is increasingly being used in the packaging of fresh fruit, ready-to-eat meals, and baked goods (Zhukov et al., 2018).

Because atmospheric O2 harms nutritional content, keeping food at low O2 tension or preventing continuous O2 input into the container is desirable for many types of food. When exposed to hydroperoxides, peroxides, and epoxides generated during the process of lipid peroxidation, tocopherols, alpha-tocopherol, and ascorbic acid are all susceptible to oxidative destruction. Except for naturally respiring fruits and vegetables and other fresh commodities, the gas atmosphere of packaged items is mostly regulated by the container type. Metal and glass containers that are properly sealed prevent gas contact between food and the surrounding environment. The permeability of the packaging material, which is largely determined by the physicochemical structure of the barrier, also influences gas diffusion in flexible packaging.

6.3.5. Light

Exposure to light can induce or hasten many negative changes in the nutritional content of meals. Ultraviolet (UV) light, with a wavelength of 2000 to 4000 A, is a kind of light with an electromagnetic vibration of 4000 to 7000 A. Shorter wavelengths of light have more noticeable catalytic properties in the visible and ultraviolet spectrums. The quantity of light and time exposed to it can have a substantial impact on discoloration and flavor faults in packaged foods. A wide range of light-screening materials has been investigated to see whether they alter food breakdown rates. Off-flavor development in fluid milk has been studied and found to be related to exposure time, light intensity, and the quantity of milk surface exposed.

6.3.6 Enzymic Reactions

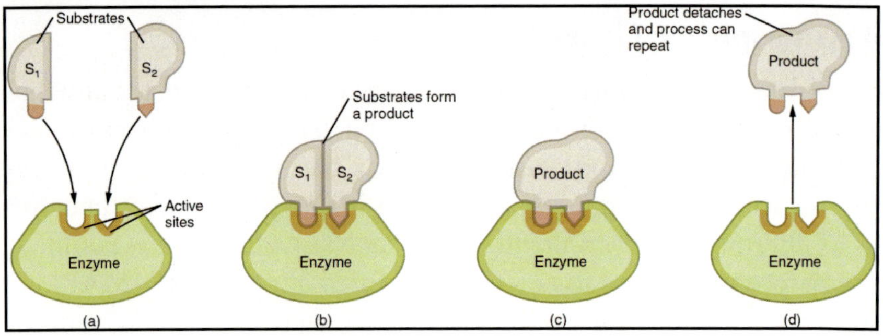

Figure 6.3: Enzymatic reactions.

Knowledge of how enzymes function is vital in the field of food packaging technology as it offers a more thorough knowledge of the repercussions of various packaging solutions. The enzymes' significance to the food processor is often defined by the circumstances present both within and outside the food. These parameters must be maintained to assure that enzymatic activity is not boosted during food preparation and storage. Temperature, pH, inhibitors of enzyme action, substrate changes, product changes, and preprocessing control are all key components in modulating enzyme activity (Weinbreck et al., 2010).

Three of these elements are particularly crucial when it comes to packing. The first is temperature, i.e., the capacity of a package to maintain a low product temperature and therefore delay enzyme action often enhances the shelf life of a product. The second critical feature is that, because enzyme activity is connected to available water, low water levels can substantially reduce enzymic activity and even affect its pattern. Finally, many O2-dependent enzyme-catalyzed reactions, such as enzymic browning in fruits and vegetables owing to phenol oxidation, include substrate modification (in particular, the entrance of O2 into a package).

When it comes to fresh produce, the faster it breathes, the shorter its shelf life; in general, the slower it breathes the longer it lasts. Peas and beans, for example, have a shorter shelf life than mature storage organ products like potatoes and onions because they respire at a faster rate.

Carbohydrates and oxygen are used to produce energy for living cells through respiration. Well-ordered biochemical pathways control and coordinate this metabolic process. It is the depletion and exhaustion of breathing reserves that causes metabolic collapse and the onset of senescence. Pathogens can enter fresh-cut fruits and vegetables if the tissues are damaged during processing, allowing cell contents to leak out. As a result, reserves are depleted and product quality suffers (Zhu et al., 2001). Tissues that don't store food, like lettuce and spinach, and immature flower crops, like broccoli, amplify this effect. The rate of respiration is reduced by temperature management, extending the useful life of the product. As a result, the shelf life of a product can be extended by using temperature control and MAP. One of the most important plant hormones is ethylene, which is responsible for the ripening and maturation of the fruit. An ether-like odor fills the air, and it's colorless. It's not uncommon for plants to produce differing amounts of ethylene, with some areas producing more than others. Ethylene's effect varies depending on the commodity, temperature, exposure time, and the amount of ethylene present. Prolonged exposure to ethylene concentrations as low as 0.1 ppm can have negative effects on a wide range of commodities. Several fruits and vegetables, such as apples, avocados, melons, and tomatoes, are particularly sensitive to ethylene.

6.3.7 Chemical Reactions

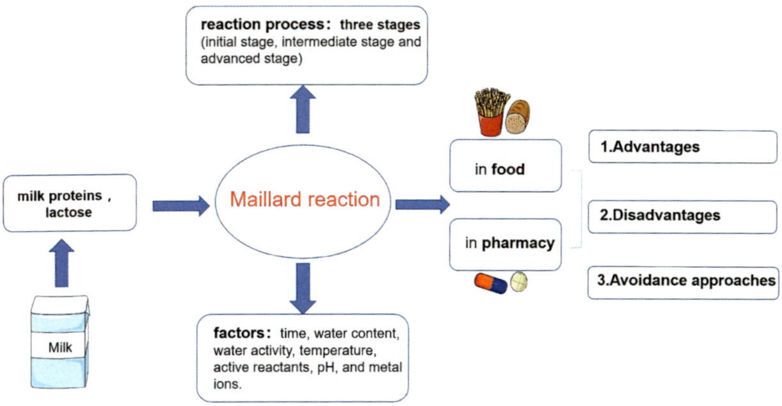

Figure 6.4: Chemical reactions. (Source: Xiang J, Liu F, et, al., 2021)

Several chemical interactions in meals may affect food quality and safety. Depending on the food and the processing or storage circumstances, different reactants or substrates can be used in various reaction types. The pace at which these chemical reactions occur is affected by light, oxygen content, temperature, and aw. In other words, packaging can influence the pace of degradation of chemical processes under particular situations (Villalobos-Delgado et al., 2019). Lipid oxidation and nonenzymic browning are two chemical reactions that contribute to the loss of sensory quality during food manufacturing and storage (NEB). Color and flavor variations are also induced by chemical reactions that occur during food manufacturing and preservation.

6.4 OXIDATION OF LIPIDS

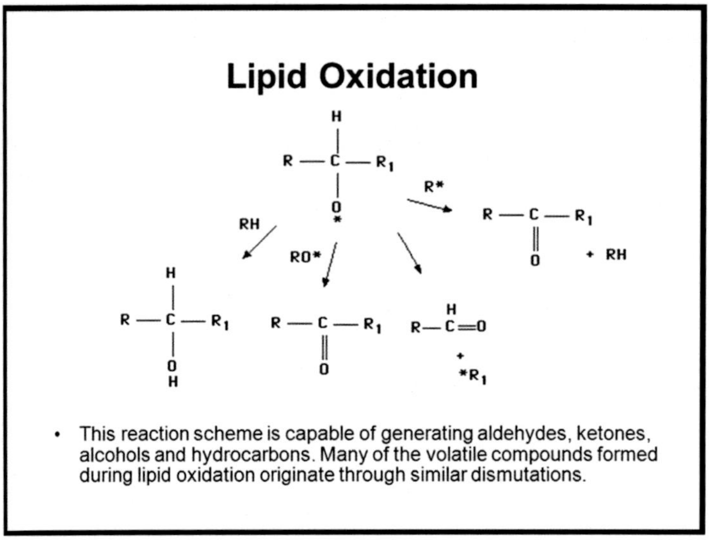

Figure 6.5: Lipid oxidation.

In the presence of free radicals, molecule O2 combines with hydrocarbons and other chemicals in a process known as "autoxidation." Due to the blazing speed of these reactions, the initiation of free radical reactions with O2 can be performed in a variety of methods. Autoxidation has long been recognized as a primary cause of food degradation and the formation of unpleasant odors and tastes. A spectrum of chemical components in food reacts with oxygen, creating changes in color, flavor, nutritional status, and, in certain situations, physical features. In certain circumstances, the effects are deleterious and shorten the product's shelf life; in others, they are essential to obtain the targeted product attributes (Amani & Gadde, 2015). Foods differ in their sensitivity to oxygen, or how much causes quality changes, as well as their avidity for oxygen, or how much they take in. Estimates of a food's maximal oxygen tolerance can be used to evaluate the oxygen permeability of packaging materials sufficient to achieve a shelf life requirement. Foods rich in fat, particularly unsaturated fats, are prone to oxidative rancidity and taste changes. Saturated fatty acids oxidize more slowly than unsaturated fatty acids.

Antioxidants can either slow or postpone the commencement of rancidity, depending on whether they are naturally present or introduced. The classic free-radical route involves the formation of free radicals in the presence of metal ion catalysts such as iron, heat, or light; photo-oxidation, which occurs when photosensitizers such as chlorophyll or myoglobin change the energetic state of oxygen; or an enzymatic route catalyzed by lipoxygenase. The subsequent breakdown of these colorless, odorless intermediates follows identical trajectories regardless of how oxygen is introduced into unsaturated fatty acids to generate hydroperoxides. The scents related to lipid oxidation are created by the breakdown products of hydroperoxides, which include aldehydes, alcohols, and ketones. Because the activation energy for both the first and second phases of the process is minimal, reducing the storage temperature does not prevent oxidative rancidity. Actions such as lowering the amount of oxygen (both dissolved and in the headspace) to less than 1 percent, eliminating elements that promote oxidation, and adding antioxidants are employed to lengthen shelf life when rancidity is a limiting concern.

Tocopherol (vitamin E), a natural antioxidant present in cocoa liquor, protects milk chocolate against rancidity. White chocolate, on the other hand, lacks the antioxidant protection afforded by cocoa liquor and is consequently prone to oxidative rancidity, particularly when exposed to light. It has a lower shelf life than milk or plain chocolate, even with the lowest barrier packaging. Removing oxygen from the pack, on the other hand, would be prohibitively costly, and the modest boost in shelf life that this adjustment would bring would not be worth the extra money (Wrona & Nerín, 2020). Packaging can be used to restrict or confine oxygen to the proper amount for a certain product. The onset of rancidity, particularly in nuts, is the limiting factor in the shelf life of snack foods. To guard against oxidation and establish a buffer against physical harm, such delicate equipment is frequently gas cleaned to eliminate oxygen and filled with 100 percent nitrogen. Metalized polyester/LDPE laminates are often employed for superior added-value products with a longer shelf life, whereas PVdC-coated OPP/LDPE laminates are commonly applied for commodity items with a short shelf life.

6.5 NONENZYMIC BROWNING

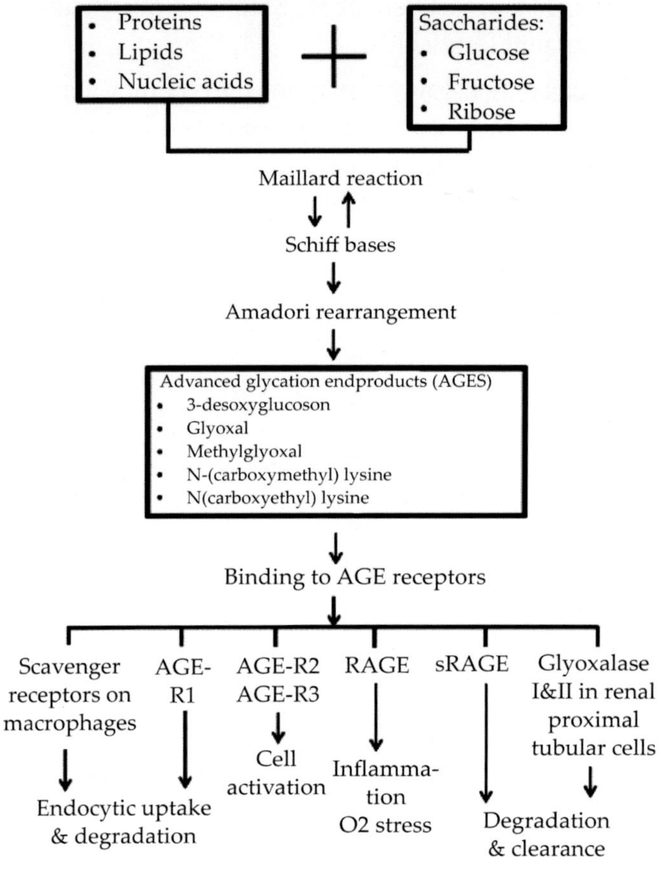

Figure 6.6: Non-enzymatic browning. (Source: Shen C-Y, Lu C-H, Wu C-H, et, al., 2020)

Browning happens in the absence of enzymes. NEB, a chemical process that occurs when dry and condensed meals are kept, is commonly seen in these foods. The Maillard reaction, often known as the NEB reaction, consists of three steps.

- In the early phases of the Maillard reaction, the amine (typically a protein or amino acid) combines with the aldehyde (usually a reducing sugar) without browning.
- This is the second form of Maillard reaction, which creates molecules that are more volatile or soluble.
- Maillard reactions result in the formation of insoluble brown polymers.

6.6 VARIATIONS IN COLOR

People's cultural, regional, and social origins all influence whether or not a specific hue is appropriate for a given meal. Despite these considerations, many food categories are only permitted if they fit inside a specified hue range (Winterhalder et al., 2015). Natural colors present in many foods, including fruits and vegetables, include chlorophyll, anthocyanins, carotenoids, flavonoids, and myoglobin.

6.7 FLAVOR CHANGES

There are different tastes to choose from. The compounding of long-chain fatty acids by enzymes is a major factor in the development of various tastes in fruits and vegetables. Off-flavors produced by these interactions can also be quite powerful. Fruity and wilted-tissue scents associated with maturation are produced by the oxidation of unsaturated fatty acids in plant tissues. High amounts of autoxidation produce aldehydes and ketones, which impart a smoky or metallic taste to foods. In contrast, it is the low quantities of these chemicals that produce the enticing aromas seen in prepared dishes. Material permeability is critical for conserving desired volatile molecules while keeping undesired components out of the environment.

6.8 NUTRITIONAL MODIFICATIONS

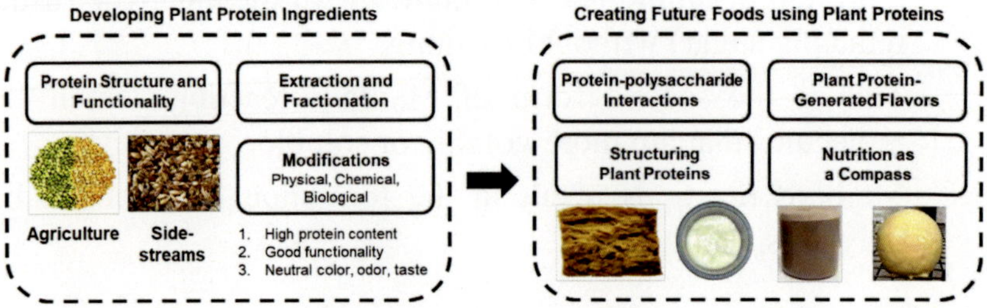

Figure 6.7: Nutritional modification. (Source: Sim SYJ, SRV A, *et, al.,* 2021)

By packing, you may change the amount of light, oxygen concentration, temperature, and aw, all of which contribute to nutritional breakdown. Because of the vast variance between nutrients, chemical variability within each class of chemicals, and the intricate interactions between these components, generalizations regarding the nutritional composition of meals cannot function.

6.9 CHANGES IN ONE'S APPEARANCE THAT ARE PURELY COSMETIC

Geometric, thermal, optical, mechanical, rheological, electrical, and hydrodynamic characteristics are a few examples of physical properties of food that may be described and measured in this way rather than by chemical means (Angourakis et al., 2015). Additionally, geometrical texture qualities can be defined as a food item's physical properties such as its form and volume. Packaging systems must take into consideration many of these physical characteristics, but the present focus is on undesired physical changes in meals that have been packaged.

6.10 MICROBIOLOGICAL CHANGES

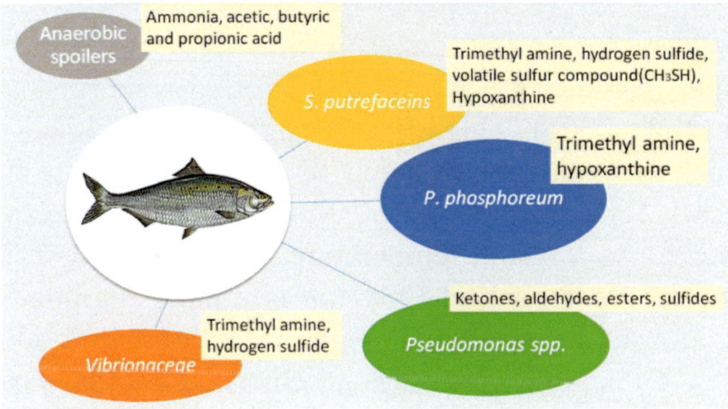

Figure 6.8: Microbial Food spoilage.

(Source: Rathod NB, Nirmal NP, Pagarkar A, et, al., 2022)

Depending on whether microorganisms are introduced as part of the necessary food preservation technique or proliferate accidentally and multiply to cause food deterioration, they can have a good or negative influence on food quality. Bacteria can't multiply, sporulate, or produce potentially harmful metabolites if their aw value falls below this threshold. Water activity influences the germination time, the length of the lag and growth rate phases, the size of the stationary population, and the subsequent death rate in each of the four key stages of the development cycle. Whether a microorganism flourishes or dies in a low environment is influenced by intrinsic characteristics that likewise influence its development at higher temperatures. These include water-binding capacities, nutritional potential, pH, Eh, and the presence of antimicrobial compounds. Solute type affects microbe growth and survival in ways unrelated to reduction (Calligaris et al., 2016). The microbial breakdown of food is influenced by factors such as temperature, oxygen, and chemical treatment. They may interact in a variety of ways to support or impede the growth of microbes. Because fruits and vegetables are living commodities, their shelf life is directly proportional to how quickly they breathe. Peas and beans have a shorter shelf life because they respire at a much faster rate than mature storage organs such as potatoes and onions. Through respiration, carbohydrates and oxygen are used to generate energy for living cells. This metabolic activity is controlled and coordinated by well-ordered biochemical pathways. The exhaustion and depletion of reserves required for breathing results in metabolic collapse and the onset of senescence. Tissues are damaged when fruits and vegetables are processed for the fresh-cut market, allowing cell contents to leak out and infections to infiltrate.

As a result, respiration accelerates, depleting reserves and degrading quality (Zhu et al., 2001). This effect is amplified by tissues that do not store food, such as lettuce and spinach, as well as immature flower crops, such as broccoli. Temperature control slows the product's respiration rate, extending its life. As a result, temperature control and MAP can be used to extend a product's shelf life. Ethylene is a plant hormone that causes fruit ripening and senescence, which is why it is commonly found in fruits and vegetables. The colorless gas has an ether-like aroma. The amount of ethylene produced by plants varies greatly, with certain parts of the plant producing more than others. The impact of ethylene varies according to the commodity, temperature, exposure time, and amount of ethylene in the mixture. A concentration of ethylene as low as 0.1 ppm can harm a wide range of products. Several fruits and vegetables, including apples, avocados, melons, and tomatoes, are extremely sensitive to ethylene.

This knowledge is critical for ensuring that a product's shelf life does not exceed its specified usage period. To achieve this goal, these components must be chosen or altered. Food poisoning pathogens can only be identified by the symptoms they cause, which can range from mild sickness to death. Infection is more likely as a result of several human diseases, and the incubation period for illness is shorter. As a result, growth must be slowed, hampered, or even restricted to some extent (Castrica et al., 2021). Some invasive diseases, such as viruses and Campylobacter, may not necessitate food cultivation. "Is the product safe?" is the first question to ask about a product's shelf life. Following that, the product's quality and commercial viability can be assessed. Food products are produced and assembled using the optimal combination of intrinsic and extrinsic components based on the starting levels of microorganisms and the conditions that kill or limit their growth rate. Predictive microbiological growth models, particularly for refrigerated foods, have proven useful in product design and packaging requirements.

6.11 EXAMPLES OF WHERE PACKAGING IS CRITICAL FOR IMPROVING MICROORGANISM SHELF LIFE

Heat processing, which eliminates bacteria, is a common method of sanitizing and increasing the shelf life of products. The quantity of heat treatment necessary is determined by the most harmful microbe present, the food's viscosity, pH, pack form, and desired shelf life. Heat, on the other hand, alters the product's texture, flavor, and appearance. As a result, there has been a shift toward minimally processed foods, which use a combination of factors to achieve the desired shelf life, such as mild heat treatment, antioxidant action, and controlled atmosphere packaging, all of which inhibit microbial growth while preserving the product's sensory and nutritional properties. Food is put in a hermetically sealed container and heated to 121°C or higher to kill all microorganisms, including *Clostridium botulinum* (Wibowo et al., 2018). It is vital to determine how long it takes for the product's coldest component to achieve the requisite temperature. The container's size and form are crucial. Because of their flat shape, retort pouches, as opposed to cylindrical cans, can reduce processing time, resulting in improved flavor and texture. Sous-vide meals are prepared in vacuum-sealed evacuated heat-stable pouches or thermoformed trays in the same way. Aseptic processing eliminates the need for heat barrier packaging, with the product and packaging sterilized separately before being blended under clean (aseptic) conditions.

To achieve the specified shelf life when using heat to produce sterility, it is necessary to use packaging that maintains sterility throughout the product's life cycle - both the packaging and pack seals must present a barrier to microbe penetration. Low temperatures have an impact on an organism's ability to develop as well as its rate of growth (Betts et al., 2004). Because specific microorganisms have evolved to thrive in cold conditions, the composition of the natural microflora will vary. Gram-positive cocci and rods, for example, are the most common bacteria in fresh milk, and if kept at high temperatures, they can sour the product. Psychotropic Gram-negative rod-shaped bacteria dominate the microbiome at low temperatures (most commonly *Pseudomonas* spp.). When the temperature is used to determine the rate at which shelf life-limiting processes operate, the function of packaging has less of an impact on shelf life because degradation occurs independently of the packaging. Frozen storage temperatures inhibit microbial growth and can result in the death of some germs, which isn't always a bad thing.

When vacuum packaging or changing air conditions are the primary shelf life-limiting variables impacting microbial growth, packing is critical for obtaining the required shelf life. To grow, Pseudomonas microbes, the most common cause of spoilage in chilled proteinaceous foods, require oxygen. Vacuum packing or oxygen-free modified atmospheres could be used to limit the growth of this species of bacteria. Other species can develop in the absence of oxygen, but at a slower rate, lengthening the time required for microbial decomposition. The gas mixture used in MAP must be selected to meet the specifications of the provided product, which is typically a mixture of oxygen, nitrogen, and carbon dioxide.

Carbon dioxide has bacteriostatic and fungistatic properties that inhibit the growth of mold and aerobic bacteria by extending their lag phase and generation time. Carbon dioxide's antibacterial effect is influenced by several factors, including microbial load, gas concentration, temperature, and packing film permeability. Because carbon dioxide is more soluble in water and forms carbonic acid at lower temperatures, the antibacterial effect is amplified; thus, temperature control is critical for maximizing the benefits of MAP and vacuum packing. Carbon dioxide, on the other hand, does not affect yeast cells but promotes the growth of lactic acid bacteria in the presence of low oxygen levels. Nitrogen is an inert gas that has no antibacterial properties. It is widely used to keep carbon dioxide-absorbing materials from collapsing in packaging and to replace oxygen in products prone to aerobic microbe multiplication. While antimicrobial packaging technologies help to prevent microbial growth, they are rarely the primary factor limiting shelf life. There are two ways to get antibacterial action.

Preservative-releasing or preservative-migrating systems contain a preservative that is designed to migrate into food and act as a preservative. Natural antimicrobial chemicals eugenol, thymol, and menthol, when combined with MAP, delayed ripening and improved fruit quality in sweet cherries and grapes by delaying softening and color loss (Serrano et al., 2008). Brassica allylisothiocyanate, a volatile molecule formed during enzyme activity, has been found to inhibit microbial growth. It has been incorporated into sheets or labels that can be applied to either the inside or outside of the packaging (Uebele et al., 2013). Permeable films, such as polyethylene or oriented polypropylene, must be applied to the pack's exterior. Wasaouro R, according to the manufacturer, is resistant to *E. coli*, *Staphylococcus aureus*, and *Bacillus subtilis*. The Japanese market sells rice cakes, bread, and delicatessen items (Potter et al., 2008). When a nonmigrating system comes into contact with the target organism, it contains or produces an antibacterial chemical. The antibacterial agent could be integrated into the packing material or applied to the surface in either case. To guarantee appropriate protection, maximum contact between food and packaging is essential.

As a result, it's ideal for vacuum-packed dinners. The antimicrobial properties and efficacy of commercially available antimicrobial packaging materials were investigated. One such technique is the use of silver zeolite, an antibacterial substance developed by AgIONTM that combines a silver active component with a non-toxic, naturally occurring inorganic clay (Timakova et al., 2020). Silver ions can be added to food polymer materials to inhibit the growth of mold, yeast, and bacteria when they come into contact with food. FoodTouch R liners (Microbeguard Corporation) are made of paper that has been infused with AgIONTM silver zeolite and can be used to extend the shelf life of both raw and cooked meals. Nisin was the most commonly used antimicrobial component in packing sheets over the previous decade, followed by food-grade acids and salts, chitosan, plant extracts, lysozyme, and lactoperoxidase. When test microorganisms are exposed to control and antimicrobial packaging films, only about two log10 decreases in microbial populations are observed, indicating that their use as part of a hurdle technology approach remains appropriate.

6.12 PHYSICAL AND PHYSICOCHEMICAL PROCESSES AFFECTING PACKAGING

Physical changes affecting shelf life may result from physical damage or physicochemical processes caused by the underlying food chemistry. Many packaging functions are taken for granted by consumers, such as product protection from environmental factors and contamination, such as dust and dirt, dehydration and rehydration, insect and rodent infestation, and product containment to prevent leakage and spillage, and physical protection action against hazards during storage and distribution. A thorough examination of the level of protection required for the product in light of the rigors of the storage and distribution chain through which it will pass is critical if it is to reach its shelf life (Turner et al., 2020). Packaging is frequently the most important factor in limiting the impact of physical degradation on product shelf life. While the risk of careless or intentional tampering cannot be considered when analyzing product shelf life, tamper-evident packaging provides a method of determining if packaging and, presumably, a preservative system have been compromised.

6.12.1 Physical Harm

Items are susceptible to vibration during transit, compressive stresses during warehouse stacking, and unexpected jolts and knocks, most notably in storage, distribution, and consumer handling, throughout their lives. The composition of the product must be able to withstand such shocks or prolonged periods of vibration; for example, emulsions must be stable enough to support vibration, and the packaging must be able to withstand and protect against such forces. Heat seals and screw caps are vulnerable areas on packs where damage resulting in leakage may result in the packaging's preservation function being lost. The outer carton shields fragile items prone to crushing, such as breakfast cereals and cookies, from physical damage and potential tampering. Whey protein films have been investigated as coverings for foods that disintegrate when handled, such as freeze-dried meals (Brody, 2009). Fruit and vegetables that are prone to bruising require special handling, and the outer packaging used for distribution must be able to withstand stacking to significant heights as well as high and fluctuating humidity. When designing packaging for this purpose, the characteristics of the commodity, such as the humidity level it can withstand, the airflow available, the rate of respiration of the product, and its sensitivity to bruising, should all be considered.

6.12.2 Insects Causing Havoc

Food infestations can be especially aggravating for customers because they frequently go unnoticed until the product is opened. Insect infestation can happen at any point after production, but it is more likely to happen during long periods of storage or transportation. Although the problem may arise through no fault of the food manufacturer, material losses can be costly, and incidents can severely harm a brand's reputation. Package bugs are classified as either infiltrators or invaders (Highland, 1984, 1991). Penetrators can penetrate one or more layers of flexible packaging. It is possible to reduce penetrator infection by using barrier materials to keep odors from escaping the container (Mullen, 1997). Mullen (1997) defines formalized a quick method for determining the effectiveness of smell barriers has been developed (Mullen, 1994). Mullen (1994) defines formalized invaders are becoming more common, and they enter packages through existing breaches caused by inadequate seals, bug holes, or mechanical damage. As a result, seals mustn't be vulnerable to insect attacks. Insects can enter through the corners of the square packaging. To reduce the risk of infection, many novel packaging methods have been developed.

6.12.3 Moisture Movement

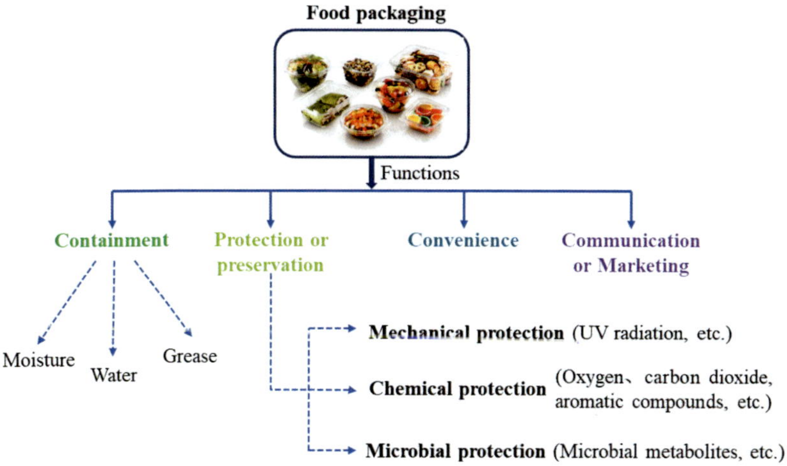

Figure 6.9: Moisture movement. (Source: Mahmud N, Islam J, et, al., 2021)

An important physical process that contributes to food spoilage is the migration of moisture, which results in the loss or gain of moisture. Hydroscopic foods, such as cereals and cookies, must be protected from moisture absorption to prevent the loss of texture. The food in breakfast cereals is best protected by the liner on the inside. Keeping moisture out of the goods while maintaining their quality is its primary responsibility. The most appropriate liner will be chosen based on the moisture permeability of the packing material and the results of shelf-life testing or a combination of breakpoint testing (holding at increasing humidities). Moisture loss during storage can be prevented by maintaining an ideal temperature and humidity level. The high value of items sold by weight makes weight loss a more pressing issue than water loss (desiccation, dehydration, or evaporation) in chilled and frozen meals.

The packaging's potential impact is demonstrated by the disappearance of chilled goods sold unwrapped from delicatessen counters, particularly prepared fresh meat, fish, and cheese. In comparison to many days to weeks, these commodities have a much shorter shelf life than packaged goods: 6 hours. Due to evaporative losses, customers prefer items that have just been loaded into the cabinet rather than those that have been sitting there for an extended period. Unwrapped items in chilled display cabinets were estimated to cost more than £5 million per year in direct evaporative losses in 1986. Thomas Swaine and John James. In stores with a high turnover rate, the average weight loss will be greater because of the frequent exposure of freshly wet surfaces to the air (Sutton et al., 2016). As a result of evaporation, fruits and vegetables stored in refrigerators will lose weight. When it comes to humidity, 97% to 98% is the ideal range for the vast majority of organisms. For practical reasons, the recommended humidity range is as a sharp product vapor pressure and velocity above the product influence the rate at which water evaporates from a given product. When a fruit or vegetable loses just 5% of its weight in water, it shrinks or wilts. MAP packaging films should have low water vapor transfer rates (WVTRs) to minimize moisture fluctuations within the package (WVTRs).

Every 10°C increase in temperature causes a corresponding increase in the amount of water needed to fill air space. To maintain the humidity inside the container at a level considered normal for the item at that temperature, food must be sealed in a container to prevent it from losing or gaining moisture. As the temperature rises, the amount of water vapor in the atmosphere does not change. Preventing moisture loss in this scenario necessitates reducing temperature fluctuations. Frozen food that gets too dry can result in grayish spots appearing on the surface of the food because cavities form in the food's outer layer. Freezer burn causes the meat's lean surfaces to become rancid, discolored, and physically altered. Materials with a low water vapor permeability are used to reduce water loss during storage and transportation. However tightly you wrap the object, dehydration will still occur because temperatures fluctuate rather than remain constant. The water inside the food package freezes as soon as it is opened. The loss of quality at the food surface is accelerated by desiccation, which increases the surface area and thus the availability of oxygen.

As a result of desiccation, which increases surface area and thus oxygen availability, the rate of quality degradation at the food's surface accelerates significantly. When it comes to retail frozen products like vegetables, removing as much air from the package as possible is difficult because they are prone to internal frost formation, especially when placed in the outer layers of display cabinets for an extended period. Laminates with an aluminum foil covering help prevent internal frost to a large extent. Sogginess is one of the most common water migration-related quality degradations. Use fat-based spreads to create a moisture barrier at the point of contact between the bread and the filling (Sewald & DeVries, 2003). Pastry and crust-based dishes like pies and pizzas are also plagued by moisture migration from fillings and toppings to the dough and crust.

Unless a suitable coloring method is used, multilayered trifles can detract from their visual appeal. Color migration across layers in multilayered trifles can detract from the visual appeal unless a suitable coloring approach is used. For example, the red streaking of the cheese is visible in pizza toppings where the cheese and salami meet. In the case of unblanched sliced vegetables in contact with dairy products, the presence of enzymes and substrates may result in taste, color, or texture issues. Originally published in 1985 in Labuza. One way to deal with these problems is to package delicate components separately and allow the consumer to mix them during use. Encasing a wafer cookie in chocolate to prevent moisture absorption, or putting an oil coating or a gelatin film over pate, can help solve similar problems.

6.12.4. Odor Detection Is Complicated By This Stumbling Block.

In some cases, a single container or trailer is used to store or transport a wide range of products. Dairy products, eggs, and fresh meat can be adversely affected by strong odors. Chocolate meals can acquire unpleasant flavors due to their high-fat content and bland flavor if they are carelessly packed and stored next to strong-smelling chemicals, such as cleaning fluids, or in stores next to highly flavored sweets, like poorly wrapped mints. The low permeability of the polymers used in vacuum packs and MAP significantly reduces but does not eliminate, the uptake of foreign scents, i.e. taints, from the air.

6.12.5. Scalability of Flavor.

If a chemical molecule in the food has a high affinity for the packaging material, it will tend to be absorbed into or adsorbed onto the material until equilibrium concentrations in the food and packaging are attained. Scalping is the term used to describe the loss of food as a result of packing. Scalping does not jeopardize the food's safety or introduce unwanted odors or flavors. However, the loss of volatile molecules responsible for the taste influences the sensory quality. Unpleasant tastes in food are frequently concealed by other flavors, particularly high notes. If the wrapping material scalps the high notes, the product's flavor is either bland or disagreeable. Scalping is affected in part by the polymer's composition and in part by the scent constituent's size, polarity, and solubility. In comparison to low-density polyethylene (LDPE), linear low-density polyethylene (LLDPE), polypropylene (PP), nylon 6 (a polyamide, PA), and polyethylene terephthalate (PET), polypropylene (PP), nylon 6 (a polyamide, PA), and polyethylene terephthalate (PET) absorbed the greatest amount of volatiles (Nielsen et al., 1992). When certain combinations of food and packaging are prone to scalping, an adequate barrier layer may help mitigate the problem. Such systems must be meticulously designed and rigorously tested to guarantee that penetration through the barrier is minimal.

6.13 MIGRATION FROM OTHER PACKAGING MATERIALS

While plastics have received the most attention in studies on food-packaging interactions, other materials such as paper, board, and cans are as harmful. Food has been wrapped in paper and cardboard for many years. They are less prone to migration than plastics due to their basic structure. Paper and cardboard packaging, on the other hand, have been linked to many food contamination incidents. Chlorophenols are capable of emitting an antibacterial odor. According to a study of a unique odor and taste in a shipment of cocoa powder, the paper sacks used to package the cocoa powder included a range of chlorophenols at concentrations of up to 520 g/kg in the paper itself and up to 40 000 g/kg in the bonded side seams. The bleaching of wood pulp for paper manufacture produced 2,4-dichlorophenol, 2,4,6-trichlorophenol, and 2,3,4,6-tetrachlorophenol, and pentachlorophenol was used as a biocide in adhesives. Pallets must be made of biocide-free wood. When fungus methylates chlorophenols, and chloroanisoles are formed (Crosby, 1981) these compounds have lower sensory thresholds than chlorophenols and can produce unpleasant musty scents (Crosby, 1981). Chlorophenols have been shown to produce chloroanisoles when present in moist paper and board substrates. Whitfield and Last (1985) have raised health concerns about using recycled paper and board for food contact (DIPNs) due to the likelihood of diisopropyl naphthalenes migration (DIPNs) (DIPNs).

These isomeric components are frequently used in the production of specialist sheets such as carbonless and thermal copy paper. According to research, DIPNs have been found to migrate into dry foods such as husked rice, wheat semolina pasta, egg pasta, and cornflour (Sturaro et al., 1995). According to a survey conducted in the United Kingdom, the use of recycled paper and board packaging materials may result in the migration of DIPNs into food. To prevent movement, it was proposed that a film or laminate be placed between the meal and the board. There was no compelling evidence that adding a film wrap to meals reduced DIPN levels. DIPNs, contrary to popular thought, may have a role in functional packaging since they have been described as possible sprout suppressants in potatoes (Lewis et al., 1997).In recent years, there has been concern over the shift away from canned lacquers and toward canned meals. BisphenolA and bisphenol-F monomers, as well as their diglycidyl ethers BADGE and BFDGE, are the most hazardous to human health. Bisphenol-A must be heated beyond the glass transition temperature of epoxy resin (105 degrees Celsius) to migrate from cans into drinks, according to the researchers. According to several assumptions, BADGE may react with dietary components during migration (Richard et al., 1999). After being subjected to tuna homogenates in water, 97 percent of BADGE could not be recognized as BADGE, its hydrolysate, or its HCl derivative.

A little amount of BADGE was converted to methyl thiol derivatives after being fed to multiple meals, but no additional molecules were identified, presumably because BADGE interacted with so many different components that the chromatographic peaks were invisible. As a result of these results, BADGE's migration into foods may have been greatly underestimated. In the 1960s, catty taint was intensively studied as another example of food-packaging interactions. A groundbreaking study discovered the production of a chemical with a cat urine-like stench (Neely, 1960). Following further investigation, it was discovered that the culprit was a sulfur derivative of an acetone-related chemical. Mesityl oxide, which is easily generated in the presence of drying agents from acetone (which had migrated from the packing), was later proven to be the source of the catty stench (Pearce et al., 1967). The taint is created when mesityl oxide reacts with hydrogen disulfide at room temperature, albeit the catty stench is scarcely detectable in trace amounts. When the tainting component is present in higher concentrations, it emits a mercaptan-like stench.

According to research on the structure–odor link of catty-smelling mercapto compounds, the volatile responsible for the catty stench is 4-methyl-4-mercaptopentan-2-one, which has a sensory threshold as low as 0.1 g/kg (ppb) on bland diets. The most current study on the prevalence of catty food taint focused on two independent complaints of a pronounced off-odor in cook-in-the-bag ham products from two different manufacturers (Franz et al., 1990). Acetone, a precursor to mesityl oxide, is present in both commodities' laminate packaging, although at amounts too low to generate catty taint (0.4 and 1 mg/m2, respectively). Following that, the source was identified as diacetone alcohol (DAA), with concentrations of 3 g/m2 and 9 mg/m2, respectively. Using an ethylene ionomer contained in the laminate packaging material, the DAA was dehydrated and transformed to mesityl oxide. Researchers discovered that eliminating mesityl oxide precursors from packing materials like DAA might aid in the prevention of catty taint in sulfur-containing products.

6.14 FACTORS AFFECTING MIGRATION FROM FOOD CONTACT MATERIALS

The amount of food contact material that migrates into food is determined by many factors. The packing material's capacity to handle a high number of potential migrants is critical. Keeping these costs low necessitates careful packaging and production. In some situations, such as when new concerns develop, it may be required to protect meals from direct contact with painted surfaces such as paper or cardboard. When preparing for potential migration hazards, it is critical to consider the amount of time food and packaging are in touch with one another. Migration is less likely if the dairy product has a short shelf life, as opposed to biscuits that have been sitting on the shelf for more than a year. When predicting how many individuals will move, it is critical to examine the main parts of a meal. As a result of a migratory component being shifted from packaging to meal, packaging concentration lowers while food concentration rises. When the concentration of the element in food and packaging is constant, an equilibrium point is attained. Packaging and food are drawn to components based on their physical qualities, such as a hydrophobic monomer-like styrene's capacity to move from one food source to another depending on its lipid content. Polystyrene migration was identified in research on cocoa powder and chocolate flakes (Linssen et al., 1991). Styrene concentrations in packing materials ranged from 320 to 1447 mg/kg, whereas those in cocoa drinks ranged from 7 to 132 mg/kg and 414 to 1447.

To balance food and packaging, a procedure akin to flavor scalping, which was previously mentioned in this chapter, is applied. The migration of paramagnetic compounds from rigid PVC in a meal simulant containing pure or mixed fatty esters indicated that the fatty esters entered and plasticized the PVC matrix (Riquet et al., 1994). To put it another way, the additive's mobility was increased as a result, as was its migration into the food simulant. The fat level in chocolate flakes with a greater fat content is g/kg. According to a sensory assessment, only the cocoa flakes were damaged by styrene. As a consequence, by examining the meal's composition, the amount of migration into food from food contact materials may be reduced. To balance food and packaging, a procedure akin to flavor scalping, which was previously mentioned in this chapter, is applied. The migration of paramagnetic compounds from rigid PVC in a meal simulant containing pure or mixed fatty esters indicated that the fatty esters entered and plasticized the PVC matrix (Riquet et al., 1994). To put it another way, the additive's mobility was increased as a result, as was its migration into the food simulant.

CHAPTER 7

STEP BY STEP DETERMINATION OF SHELF LIFE OF FOOD

CONTENTS

7.1 INTRODUCTION

Food shelf life refers to how long a product may be supplied to customers before deteriorating quality or being declared unfit for human consumption. However, because the bulk of meals declines in quality over time, it is vital to determine how long these things will retain their economic value (Chang et al., 2022). The purpose of shelf-life testing is to provide an objective assessment of the qualitative attributes of the food as well as the point of failure. This failure occurs when the client is dissatisfied with the physical, chemical, microbiological, or sensory aspects of the meal. Shelf-life testing is a complex, time-consuming, and costly activity. Throughout the test, quality indicators must be re-evaluated regularly (Timakova et al., 2020).

Typically, such testing will last at least as long as the product's allowed shelf life. In certain cases, this might take a year or more. A three-month product development cycle certainly does not allow for such a lengthy laboratory test. The required sample size and analysis costs may be significant.

The vast majority of shelf-life tests are performed at the microbiological or sensory level. While this is true for some foods, the product designer can use a degradation index over time to assess the sensory quality of the product. Changes in moisture and water activity include the presence of ascorbic acid and malonaldehyde in decomposing fish, as well as changes in the water holding capacity and peroxide value (Sutton et al., 2016).

Even in the absence of an analytical shelf-life test index, many solutions have been proposed to reduce the testing burden. Statistically, experimental designs that use fewer samples but yet produce reliable and valid results may save money.

Several strategies have been developed to improve the speed with which shelf-life assessments are performed. T. The fundamental ideas underlying these types of tests were developed by Labuza at the University of Minnesota. Accelerated shelf-life testing is keeping food in a temperature, humidity, atmospheric, or light-controlled environment where one or more storage parameters are kept at a higher level than usual (Dawson et al., 2018). This accelerates the deterioration process, resulting in a faster period of product failure than would otherwise be the case. Because the Arrhenius equation may be used to calculate the influence of external factors on deterioration, the amount of acceleration and the product's true shelf life under normal conditions can be predicted (Sewald & DeVries, 2003). Accelerated shelf-life testing, according to Labuza, requires the user to obtain at least one analytical measurement related to product quality and then numerically estimate the deterioration kinetics.

The application of mathematical models based on functions that characterize a product's degradation processes allows for system analysis and adjustment. To anticipate a product's storage life, a variety of scenarios and failure criteria can be used. Mathematical modeling is based on many hypotheses. If any of these assumptions are incorrect, the model loses its significance (De Corato, 2020). Typically, models incorporate correction variables to account for any situations that may have a significant impact on shelf life.

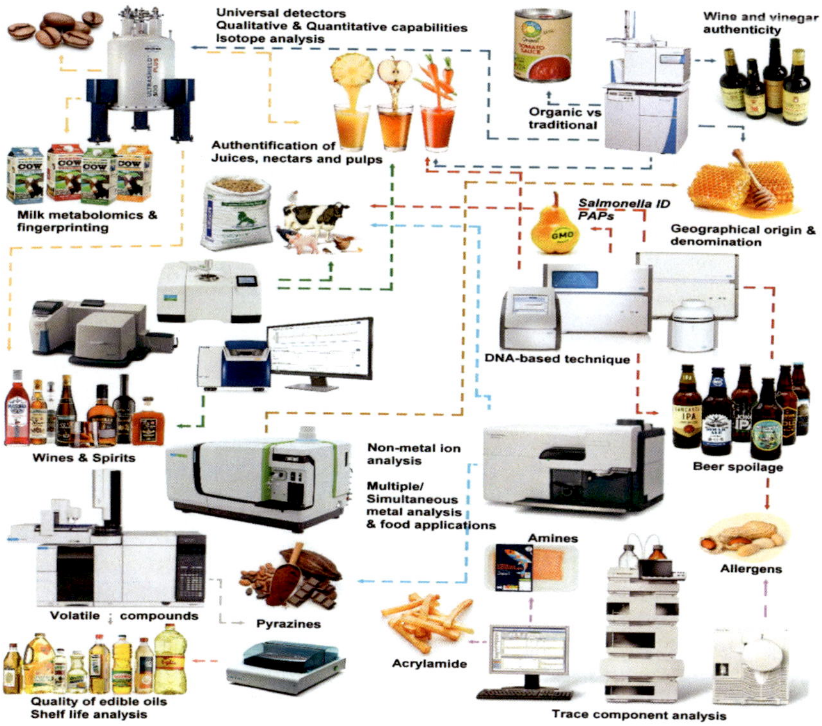

Figure 7.1: Some of the methods used to determine the shelf life of food.

(Source: Artavia G, Cortés-Herrera C., et, al., 2021)

A shelf-life study is a method of determining the predicted shelf life of a food product without sacrificing quality. Each product needs a certain amount of study. There are two primary ways, as a general rule:

- **The direct method**: This is the most commonly used method. The goal is to preserve the product beyond its stated shelf life by storing it under specified conditions and periodically analyzing it for signs of degradation. Each product needs its technique. The next sections detail the procedures to be followed as well as the types of choices to be made (Sumner & Jenson, 2011).
- **Indirect method**: This method predicts shelf life by using accelerated storage and/or predictive microbiological modeling.

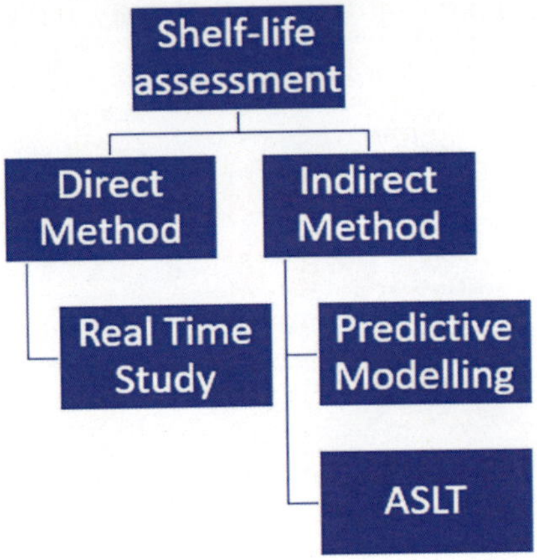

Figure 7.2: Shelf life assessment.

7.2 STEP BY STEP DETERMINATION OF SHELF LIFE BY THE DIRECT METHOD

STEP 1: Determine what is causing the food to deteriorate or become unsafe. Every product's shelf life is influenced by a variety of factors. Consider the following lists of potential reasons why a product's quality or safety can deteriorate as a starting point. At the same time, identify the components that contribute to shelf life extension. Consider the entire process, from the procurement of ingredients and packaging materials to the consumer's final use (Sarkar et al., 2022). Because of the higher temperatures, certain goods deteriorate faster in the summer than in the winter.

PRODUCT SPOILAGE: Raw materials: The final product is affected by raw material quality, uniformity, contamination level, and storage. Cosmetics for the face include: The substances you use and their interactions influence the type and quantity of spoilage organisms that can grow. Changes in the components or chemicals themselves, as well as changes in their quantities, can have an impact on shelf life. Water activity refers to the amount of water in a meal that bacteria may utilize (El-Anany et al., 2009). Water is required for the growth of microorganisms. Water activity can be reduced by adding salt, sugar, and other substances. Jam is a moist meal, but bacteria can only use a tiny portion of the water due to the high sugar content.

The pH scale is used to determine a solution's acidity and alkalinity. The pH of a meal determines which bacteria may survive and develop in it. Fresh mussels, for example, have a limited shelf life, but mussels marinated in vinegar (acidic) marinades have a significantly longer shelf life because the acidic environment inhibits the growth of the majority of bacteria. Breathable air supply: It is possible to extend the shelf life of food by eliminating the surrounding air, using vacuum packing, or using modified environment packaging. However, in industrial systems, certain bacteria that thrive in low-oxygen settings must also be controlled (Roccato et al., 2017).

Unlike bacteria that produce slime, botulism, and listeriosis, which do not need oxygen to survive, these packing strategies kill or inhibit the growth of decomposing organisms that do. Chemopreservatives are chemical preservatives. When applied appropriately, preservatives aid in the management of microbial development. Sulfites, for example, delay the decomposition of dried fruits and keep them from browning. The Food Standards Code of Australia and New Zealand govern the types and quantities of preservatives that are authorized.

The process-related to spoilage: Processing: Processes include mixing, salting, smoking, fermenting, heating, cooling, and chilling, as well as dehydration, freezing, and heat sterilization. When it comes to shelf life, the end product's manufacturing process can have a considerable influence (Fan et al., 2009). Because of the additional heat treatment, UHT milk, for example, is sterile and has a longer shelf life than pasteurized milk. Ingredient quantities, quality, and production procedures must all be consistent in order to consistently produce a product with the same look, flavor, and shelf life, among other attributes. Using a recipe or flow chart that describes exactly what has to be done at each stage, including timings and temperatures, will assist in ensuring this. You can't take advantage of someone else's shelf life unless you're convinced that their entire operation (including cultivation, production, distribution, and sales) is similar to yours. Microorganisms that cause food-borne illness and deterioration may tolerate even minor temperature or temporal changes. When a product is transported, sold, or stored in a home environment, it must be properly wrapped to avoid contamination. Packaging can be a source of infection if it is not designed and managed properly (Pratiwi et al., 2015). Temperature management during storage is critical because it can prevent the formation of bacteria that contaminate and damage food. When storing food, the existence of rodents, birds, and insects, as well as other components that might contaminate the food, should all be considered. Temperature data recorders can be an important instrument for guaranteeing precise management since variable temperatures have the potential to impair the shelf life of perishable commodities. At this point, you should have identified all of the possible ways the product might decline, as well as the underlying causes.

STEP 2: Which tests should be carried out.

To ensure the product's safety and quality, you must do the necessary tests. There are a few tests that don't apply to every product. However, raw fermented meats would not be tested for lactic acid bacteria. Testing for foodborne illness organisms such as Listeria may be necessary to confirm the product's safety. For any laboratory testing, IANZ accreditation is required (Peleg et al., 2017). Storage temperature stability data from storage experiments performed on your behalf by a laboratory must be reviewed.

In general, tests can be divided into the following four categories:

- SENSORY EVALUATION

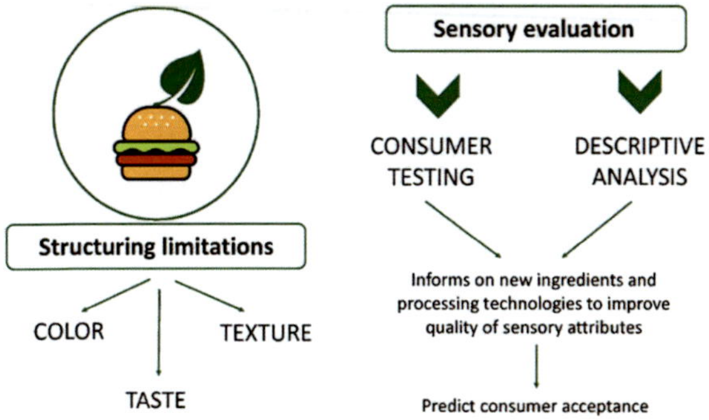

Figure 7.3: Sensory evaluation. (Source: Fiorentini M, Kinchla AJ, et, al., 2020)

The sight, smell, taste, and feel of food are all included in the sensory assessment. It may be used to monitor and record large changes over time, making it useful for calculating food shelf life. Foods should be inspected at the point where they will be kept and consumed to verify that they are safe to eat. This should ideally be done by a certified panel using well-established evaluation techniques. Before bringing a tasting panel into your home, always double-check the food's safety (Omotilewa et al., 2018). If at all feasible, freeze samples before the trial begins to avoid the meal's texture or other aspects from being affected. These can then be utilized as a starting point for additional testing. If the meal cannot be frozen, a newly prepared sample should be utilized instead.

- MICROBIOLOGICAL.

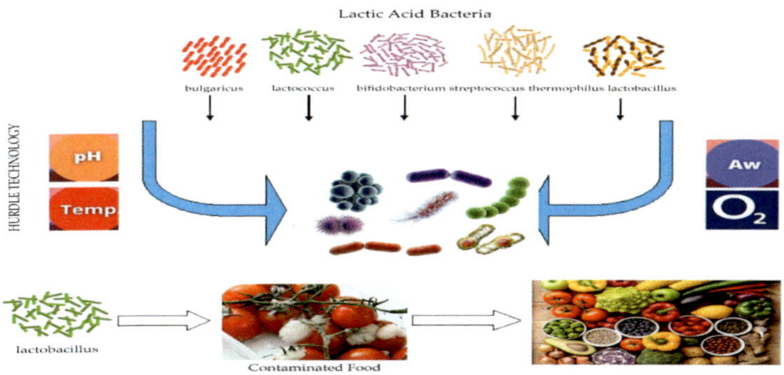

Figure 7.4: Microbiological. (Source: Ibrahim SA, Ayivi RD., *et, al.,* 2021)

These tests may be used to determine both the nutritional value and the safety of food. Several tests may be used to evaluate the presence and type of spoilage organisms (yeasts, molds, and bacteria) over time. If a certain food item is missing from the table, arrange it according to your best judgment. To maintain food safety, it is vital to detect any organisms that may cause food poisoning (Nagi et al., 2012). The amount of testing required varies depending on the product. Microbiological guidelines and recommendations describe the types and numbers of organisms considered safe or harmful in food. Food testing in New Zealand must meet the following requirements: Food Microbiological Limits. A set of criteria was used to assess microbiological recommendations (O'Brien & Bentley, 2015). Microbiology guidelines for ready-to-eat foods. Taking into consideration the other criteria mentioned above, the food sector can better estimate the right microbe levels. The industry, according to the NZFSA, should develop its own, more stringent microbiological requirements. Several European countries are developing microbiological criteria for *L. monocytogenes* that have outlived their usefulness. This guide includes a section on estimating the shelf life of foods that support the growth of *L. monocytogenes*.

CHEMICAL.

Chemical analysis can detect changes in a product's quality throughout its shelf life. Chemical testing employs four types of chemical instrumentation: pH meters, headspace analyzers, free fatty acid meters, and total volatile nitrogen meters.

- PHYSICAL

Texture testing, packing inspections, "journey tests," and estimating the ideal, optimal, and average retail situations are among them. The use of a "travel test" can help discover potential transportation and handling dangers. It necessitates delivering and keeping things in the order that they were meant to be supplied. Tests must be carried out along the supply chain, not only at the end. A data logger can be used to record the temperature at regular intervals for further examination (Nagi et al., 2012). The projected transit temperatures to retail outlets, commercial control, consumer purchase, consumer transfer and storage, and so on should all be included in the experimental design.

STEP 3: Plan the shelf life study.

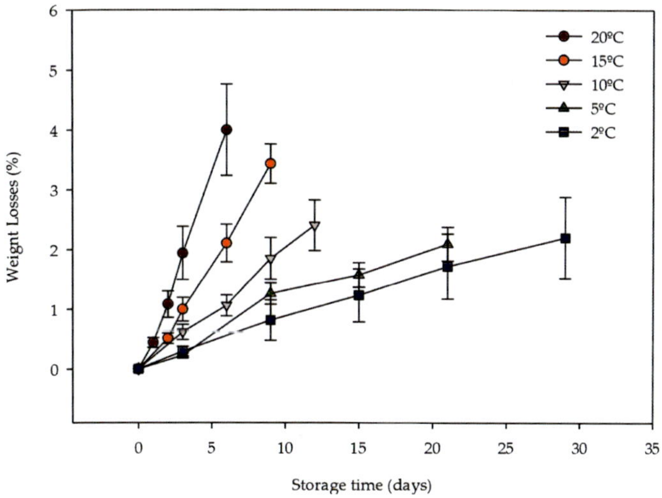

Figure 7.5: Shelf life study(Source: Torres-Sánchez R, Martínez-Zafra MT, et al., 2020)

The following step is to perform shelf life investigations. When developing a complete shelf life study approach, keep the following in mind:

- What sorts of tests are required?
- How long will the study endure, and how often will exams be given? Make a plan that includes the example's precise dates. It is best to collect samples at the start, end, and three places in between. To validate the endpoint selection, further sampling beyond the objective is necessary.
- At each checkpoint, how many samples will be evaluated? At least three packets of goods should be evaluated at each sample.
- How many specimens will be gathered throughout the investigation?
- When will the study be done? It is unknown when the probe will be finished. It's preferable to do it during the summer when issues are more likely to arise. The experiment should be repeated several times to account for product variations (Fan et al., 2009). The final product should be similar to the prototype in terms of product, technique, and packaging. Keep track of everything you use or do since this will help you assess the repercussions. Investigations are now being planned and prepared in great detail.

7.3 HOW TO CONDUCT A SHELF LIFE STUDY

There are several ways of determining shelf life, and no single approach is widely recognized. The Food Safety Authority of Ireland (FSAI) and the Government of New Zealand have both established regulated criteria for estimating food shelf life, which can be found here. To conduct shelf life research in a systematic manner, the 10 stages outlined below must be performed exactly.

Before any study begins, the end goal of shelf life research should be specified. A new product, a modified recipe, or a reexamination of new packaging are all examples of events that might prompt the start of a shelf life inquiry. Determining the degradation process is critical since the shelf life of various foods varies based on the kind of food (El-Anany et al., 2009). When there is a high concentration of degrading microorganisms in the environment, cold things lose their shelf life. This involves, among other things, the oxidation of fat and the deterioration of vitamins in juices and bread, as well as the retrogradation or aging of starch in slices of bread and other baked items.

A range of tests can be done during shelf life inspection, including microbiological inquiry, chemical analysis (such as fat oxidation and vitamin degradation), physical testing (such as color or viscosity), and sensory assessment. It is necessary to create a link between changes in sensory perception and the value of the analytical measurement.

The formation of reactive peroxide, free fatty acid, or thiobarbituric acid molecules, among other things, can be measured in a lipid oxidation test. Sensory assessment may be assessed using tests such as discrimination, description, and acceptability. When storing items, it is important to keep in mind the importance of factors such as temperature, relative humidity, and light (De Corato, 2020). The circumstances under which goods are stored might range from optimal to dreadful. Depending on the application, static or dynamic variables can be utilized to replicate real product exposure throughout storage, distribution, and retail activities.

Determine how frequently and when goods with a limited shelf life are evaluated (seven to ten days). When it comes to items that have a short shelf life, it may be essential to replace them every day or every two days (seven to ten days). If an item has a moderate shelf life (three weeks) or a long shelf life (over a year), testing can be done at the beginning and end of the shelf life, as well as two to three times in between and once after the shelf life has expired (one year). Determine if the test and control samples are suitable for your experiment - this is a vital step.

Determine which components, processes, and containers will be employed in the shelf life experiment. Test samples should be drawn from the same batch as the batch being tested, with enough samples saved for a second or third test run. A newly manufactured product can be used as a sensory control, and samples can be kept cold or frozen to ensure that the product does not deteriorate over time due to the production process (Dawson et al., 2018). Conduct an assessment to determine the shelf life of the samples. Continue storing samples in research settings and evaluating them regularly.

Conduct a comprehensive review of the results. In the absence of restrictions, manufacturers must design their termination criteria based, among other things, on microbiological, chemical, or sensory variables. Examine and assess the results (legal or voluntary). The expiration date of a valid sensory score or an acceptable set of analytical data is normally given as the last day on which they are available. Because of the worst-case manufacturing and storage conditions, the expected shelf life may be overly optimistic (O'Brien & Bentley, 2015). After a product is released, it should be monitored and evaluated to verify that it is still in excellent working order, and the expiration date should be modified accordingly.

STEP 4: Investigate the product's shelf life.

Throughout the investigation, samples should be preserved under the same conditions as typical production samples, from fabrication through consumption. If feasible, samples should be kept at the same temperature and relative humidity during their entire storage period. Frequently, these must be validated and documented. At the dates stated in Step 2, samples are selected and evaluated by the processes outlined in Step 1. Keep a record of the test results and anything else that may be beneficial in the future (Chang et al., 2022).

STEP 5: Determine the product's shelf life.

Eventually, the product deteriorates to the point that it no longer meets the quality standard. Using all of the information you've gathered, determine the maximum duration the product may be stored while maintaining acceptable quality and safety. Maximum storage times for quality and safety may range. The shelf life of a product should be determined by which has the shortest period. Examine the exam results and repeat any that appear unreasonable. Check that the components, quality, and processes are all the same in all batches to ensure consistency. Find and correct the cause of the variation before resampling and testing.

After estimating the shelf life under ideal storage circumstances, you must consider the "real world," where storage conditions may change and items may be misused. The shelf life of your product should be practical, not ideal, and should contain a safety buffer. Product abuse can be reduced by defining and controlling the product's circulation. You've calculated the useful life (Calligaris et al., 2016).

STEP 6: Take note of the expiration date.

Figure 7.6: Expiry date.

Acidity, taste loss, the amount of decomposing organisms, and other characteristics identified as important by shelf-life studies should all be tested in samples. Samples can be collected at many points throughout the distribution and retail chain. To be safe, the preliminary shelf life should be enhanced at all times. Furthermore, any modifications to the manufacturing or processing environment demand a shelf life analysis. Finding the limiting elements in the shelf life research, altering them, and repeating the study can be utilized to prolong the shelf life (Angourakis et al., 2015). Investigating consumer complaints about things that do not perform as advertised may reveal a recurring problem and the need to recalculate the shelf life. The information gathered during the design and execution of the shelf life research may be used to analyze consumer complaints, manage manufacturing and distribution issues, and investigate the product's shelf life. All test findings must be documented and stored in a safe yet immediately accessible area.

According to the manufacturer, the expiration date of a product indicates the last day on which it will be of the highest quality. A product's expiration date can be found on the label. The expiration dates for food and medicine are opposed to one another. Under federal law, only infant formula must be labeled with an expiration date. When this occurs, it usually indicates that the food will be at its peak in terms of taste and texture up until the point at which it is consumed. Expiration dates are mandated by federal law for both prescription and over-the-counter medications, but they do not represent a safety guarantee. Since then, the product's safety and effectiveness have been confirmed through the use of the manufacturer's testing procedures and protocols (Omotilewa et al., 2018). Whenever a product is intended to be sold or used (or even eaten), a date may be printed or imprinted on the package or container. They all have distinct meanings, but only the "do not use after" date specifies that the product should be thrown away after that date because it may be harmful, ineffective, or both if used after the specified date.

In the retail industry, specifying a product's "sell-by" date is an important part of the process. When the quality of a product begins to deteriorate, the use-by date serves as a warning to the public that the product should be discarded. Even after a product's "best by" date has passed, its flavor or texture may begin to deteriorate, with the exception of infant formula. While the United States Department of Agriculture (USDA) provides some guidance on proper food date labeling, it does not provide much assistance. It is estimated that 30 per cent of the nation's food supply is "lost or wasted," in part because people throw away perfectly edible food. A better indicator of whether or not a portion of food will be accepted is a smell or taste test rather than a label (Amani & Gadde, 2015).

7.4 Step by Step Determination Using Indirect Method

Two indirect methods are used in determining the shelf life of foods. These are:

- Accelerated shelf-life testing.
- Predictive microbiology.

7.4.1 Accelerated Shelf-life Testing (ASLT)

To accelerate product breakdown without interfering with normal storage conditions, the ASLT program was created. There must be no further alterations made during any ASLT when testing conditions are meant to accelerate the aging of the specimen. It is particularly advantageous to utilize ASLTs to estimate the shelf life of long-lasting, ambient-stable foods such as confections, as this approach is quite accurate. ASLTs, for example, cannot be employed on materials that have microbial growth on the surface (Timakova et al., 2020). The major purpose of this research is to accelerate the rate at which physicochemical changes occur. When it comes to materials that deteriorate as a result of microbial proliferation, several relevant tests may be used to predict growth. Mizrahi delves into the underlying concepts of ASLT in great detail (2000 and 2004). The use of ASLTs has several disadvantages, even though they may be quite beneficial in some situations. The first step is to have a full understanding of the product, which will allow you to design faster testing that will reliably forecast shelf life. Synchronization of ASLTs is required to compare the rate of change of certain attributes in new products to the known shelf life of similar products.

The storage conditions that cause the product to degrade at a quicker rate are dictated by the product's quality. Normal storage conditions eliminate the need for accelerated testing under conditions that could cause the product to become incompatible with the storage environment. Dark chocolate will have a temperature limit of 30 degrees Celsius and milk chocolate will have a temperature limit of 24 degrees Celsius due to the melting of fat at higher temperatures. ASLT parameters should be determined concerning product stability as well as anticipated environmental conditions, according to the manufacturer. In comparison to Schaal oven studies, incubations in ASL research are far more advanced and well-controlled than those in Schaal oven experiments. Investigations into ASL provide a fundamental context for what laboratories perform daily: measure the rate at which lipids oxidize (Sutton et al., 2016). By putting samples into matching containers and placing them in incubators, it is possible to obtain exact temperature and vibration control, as well as precise atmospheres and environments. Samples can be exposed to specific wavelengths of light or kept in complete darkness, depending on the application. Samples are obtained at regular intervals and evaluated using a range of chemical and physical methods to track the progression of oxidation in the environment. Fundamental oxidation reactions, as well as the stability of various oils or foods, food processing and formulation, catalyst and antioxidant evaluation, genetic modification guidance, and determining "sell by" dates, among other practical and fundamental research questions, can all be investigated using the same approach.

Despite its limitations, learning ASL can be beneficial. While not as precise as a standard Schaal oven test, the ASL is based on the assumption that at elevated temperatures oxidation pathways, catalysts, and antioxidant effects are identical to those found at room temperature. This allows for much more rapid oxidation calculations than would otherwise be possible. In reality, lipolysis is a series of reactions, each of which has its activation energy, and so isn't a single reaction at all (Peleg et al., 2017). Lipolysis may be thought of as a series of reactions since each one has its activation energy. The presence of a wide variety of catalysts in fats and oils has the potential to stimulate other types of autoxidation reactions that are distinct from the fundamental autoxidation reactions and have different activation energies as well as different types of improved products (Sewald & DeVries, 2003). Because the presence and balance of specific pathways and catalysts must change as the temperature rises, it is impossible to anticipate the presence and balance of the same lipid oxidation products at all temperatures. The results of ASL experiments under catalytic conditions may reveal how oxidation rates and product distributions change; however, they should be used with caution, if not outright scepticism, when attempting to predict how oxidation would occur under less catalytic conditions.

The image of oxidation that is formed is influenced by the product(s) under investigation, the frequency with which studies are conducted, and the interpretation of results as a result of specific chemical interactions, whether broad or narrow in scope. The shelf-life evaluation differs depending on whether only hydroperoxides are collected and reach a peak after ten days before degrading or whether only hexanal is collected and continues to accumulate for more than twenty days before degrading. It is also possible to exaggerate the oxidation levels of hexanal when other C9 products such as nominal and decadienal (which develop at higher temperatures) are detected; nevertheless, the shelf life may be inflated since the oxidation processes have been changed as a result of the measurement (Sumner & Jenson, 2011). All of these observations have been made in the laboratory. It is necessary to employ as many analytical tools as possible to get as much information as possible to get the most out of the information you have gathered.

Furthermore, if at all possible, analyses should be able to discriminate between individual items and whole classes. We have seen a wide variety of aldehyde and alkane distributions in both volatile and nonvolatile compounds, even at extremely low aldehyde concentrations. This has been reported in both volatile and nonvolatile substances. When total carbonyls were analyzed, it was discovered that there were no changes in the product distribution channels. Although ASL is designed to reflect long-term preservation, it is necessary to review samples more frequently than once a week or once a month to discover the precise patterns of change.

7.4.2 Predictive microbiology

The development of mathematical models that can forecast microbial activity in food contexts is a primary objective of predictive microbiology in foods, a branch of microbiology. Predictive models have been present since the early 1900s, but advances in computer software have expedited their expansion in recent decades (Sarkar et al., 2022).

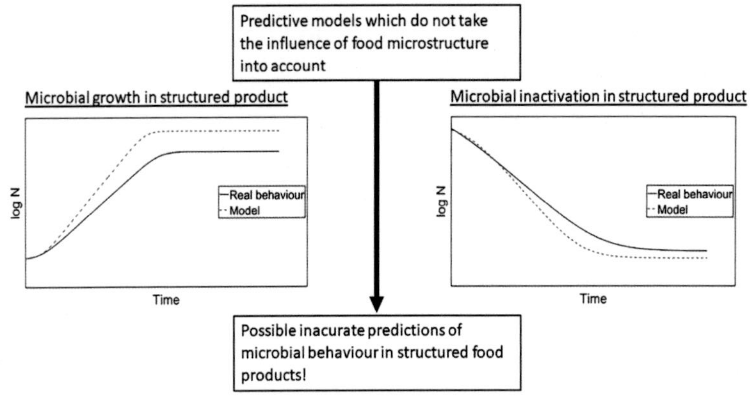

Figure 7.7: Predictive microbiology. (Source: Verheyen D, Van Impe JFM. *et, al.,* 2019)

Future predictive microbiologists must be conversant with mathematical and modeling concepts, which should be taught to them before they start their careers. They should also be well-versed in food microbiology, which they should be taught before beginning their jobs. Predictive models can utilize characteristics like cell physiological state or microorganism interactions to estimate bacterial growth, inactivation, and the chance of growth in meals. Because of its capacity to respond quickly to specific situations, predictive models have emerged as a helpful tool for assisting with food safety and quality decisions (Roccato et al., 2017). In order to increase the performance of self-control systems, HACCP and food safety risk-based metrics have been added as useful components, such as predictive models. It is presently at the forefront of national and international food safety standards to promote the advancement of Quantitative Microbial Risk Assessment (QMRA) research, which is solidly backed by scientific methodology. Predictive microbiology is a new field that is rapidly expanding to help ensure the safety and quality of the food we consume.

Microbiological testing in laboratories is commonly used to determine food safety and product shelf life. Food-borne microbe formation, survival, and inactivation, on the other hand, are all common occurrences. Microbiological models, which investigate how various environmental components interact with bacteria, can be used to forecast how long bacteria will live, develop, and stay inactive. Predictive microbiological models in the food sector can aid in the evaluation of new formulations for safety and stability, as well as the discovery of formulations with a long shelf life. When a product's shelf life has been calculated but the product is later subjected to a minor technique or formulation change, predictive microbiological models can be quite useful (either planned or unplanned through loss of process control). After that, a microbiological model that predicts future changes may be used to assess the product's safety and shelf life (Pratiwi et al., 2015). Using microbiological simulations, product developers can determine the barrier combinations that will allow them to reach the specified shelf life. These specific characteristics may be tested experimentally to save money on time-consuming and expensive research.

Certain constraints apply to the model's construction.

When developing predictive microbiological models, it is believed that microbial reactions would remain consistent across time. Predictive models can provide a cost-effective method for removing microbiological testing for the goal of determining shelf life, albeit the model's predictions may be wrong due to changes in bacteria reactivity and the growing environment (Peleg et al., 2017). As a result of the study's findings, predictive microbiological models are unable to effectively anticipate the survival, development, or inactivation of illnesses in food products. To ensure food safety, it is critical to assess predictive microbiological models before depending on them to aid in making food safety decisions. Validation in computer science is the process of comparing a model's predictions to data that was not used in the model's design.

CHAPTER 8

PREDICTING MICROBIAL SHELF LIFE

CONTENTS

What do we mean when we say "shelf-life"? It's the period after production when a food or beverage product is still safe to eat or drink. Isn't that a simple definition? However, when it comes to developing shelf-life protocols, the food and beverage industry faces numerous challenges. It's a difficult task. Products can experience a wide range of temperatures as they transition from one location of the cold chain to the next, from production to buyer storage. It's difficult to figure out what these conditions are and how long they've been there. However, it must be done to deliver a useful shelf-life protocol (Hedegaard & Skibsted, 2013).

For manufacturers to follow, retailers have frequently established their own organization's shelf-life protocols.

They seek information like, "How long will the item be out of the cold when bought?" when designing them. and "At what atmospheric conditions will the item be kept during the buying process and in the customer's refrigerator?" Frequently, retailer guidelines vary, resulting in a manufacturer providing comparable products to various customers while obeying various testing protocols.

This raises the question of whether these discrepancies in shelf-life guidelines have a massive effect on the shelf-life designated to a product. Will spoilage bacteria behave similarly? How much faster can they develop and spoil the goods before it's too late? What is the potential for a safety hazard? We can answer these questions with the help of predictive modeling.

8.1 WHAT IS PREDICTIVE MODELING?

Predictive modeling is a technique for determining a microorganism's survival and/or growth under a variety of conditions such as temperature, pH, and water activity (aw). The models are created with computer-based software applications and can be employed to rapidly assess the impact of various storage regimes on product microbiological levels (Martins et al., 2008). It's also very useful during the product development phase, as it allows for a quick assessment of the implications of recipe changes, such as the effect of reducing salt on Listeria. All in all, it can be used to estimate a product's probable shelf life, both in terms of quality and safety.

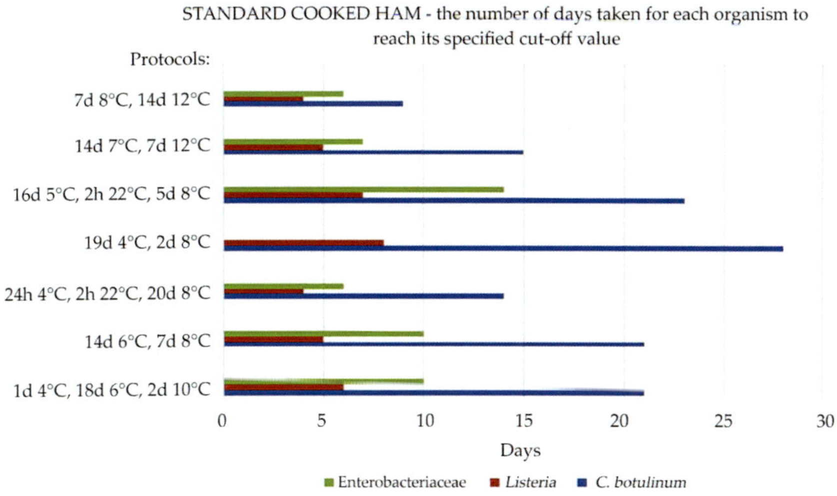

Figure 8.1: Potentially achievable shelf-life in standard cooked ham – the number of days taken for each organism to reach its specified cut-off value.

Modeling can be used in conjunction with real-life sample testing. In challenge testing, relevant microorganisms are deliberately inoculated into a product and their growth is monitored over time. This method takes into account all variables that may have an impact on growth. Predictive modeling, on the other hand, usually only takes into account three variables: pH, aw, and temperature. The following are some of the unaccounted for factors:

- Antimicrobials that are naturally present (e.g. allicin in garlic)
- Structural differences in a food sample
- The evolution of aw and pH over time, and
- Organisms that compete

8.2 WHAT IS PREDICTIVE MODELING AND HOW DOES IT WORK?

The models are created using microbial growth curves generated in the lab. Microbiological growth media (broths) with various intrinsic parameters (e.g., pH and salt levels) are cultured with the applicable organism or concoction of organisms and nurtured at various temperatures (Lindström et al., 2006). The data is then fed into mathematical equations, which are then used to build models that predict how organisms will grow in situations similar to those of a specific product. As a consequence, we can use this information to forecast how long this product will last.

8.2.1 What Predictive Modeling Can Tell Us About Shelf-Life Protocols?

We used this method to look at how 7 distinct protocols for determining shelf life showed differences in the growth of various microorganisms. For the experiment, cooked meat and raw meat were used. The goal was to figure out how minor variations in shelf-life protocols affected shelf-life.

Two approaches were used to accomplish this. One to assess the product's safety (by measuring the development of two pathogenic species: *Listeria* and *C. botulinum*) and the other to evaluate the product's quality (by assessing the degree of spoilage organism Enterobacteriaceae).

The 'end of shelf-life for each organism was defined as the time at which *C. botulinum* increased by 0.5 log cfu/g, Listeria increased by 10 to 100 cfu/g, or Enterobacteriaceae reached 106 cfu/g. The item was then deemed unsafe or spoiled at this point (Liu et al., 2018).

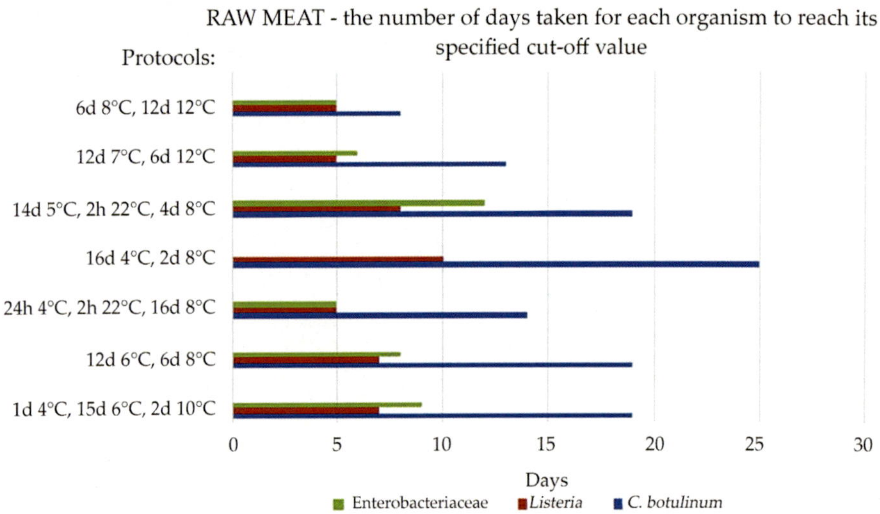

Figure 8.2: Potentially achievable shelf-life in raw meat – the number of days taken for each organism to reach its specified cut-off value

8.2.2 Predictive Microbiology and Shelf-life

Overall, predictive microbiology models aim to quantify the effects of intrinsic, extrinsic, and/or processing factors on the subsequent microbial spread in food products or food model frameworks, such as a buffer system. These models rely on the ability to extrapolate the subsequent microbial proliferation for permutations that are not only tested but also fall within the experimental design's scope. As a result, predictive microbiology can be thought of as a useful tool for examining and summarizing the impact of various circumstances (within food preparation and processing) on microbial ecology (Kanatt, 2020).

Mafart introduced a historical context of modeling progress in the area of predictive microbiology in 2005. Per this, the first initiatives date back to the 1920s, when the Arrhenius equation or the Bigelow model were used to defining the heat vulnerability and tolerance of bacteria. Nonetheless, the discipline's precepts and goals did not emerge until the early nineties, followed by the formation and summary of microbial models, as well as the creation of reference lists and other software solutions.

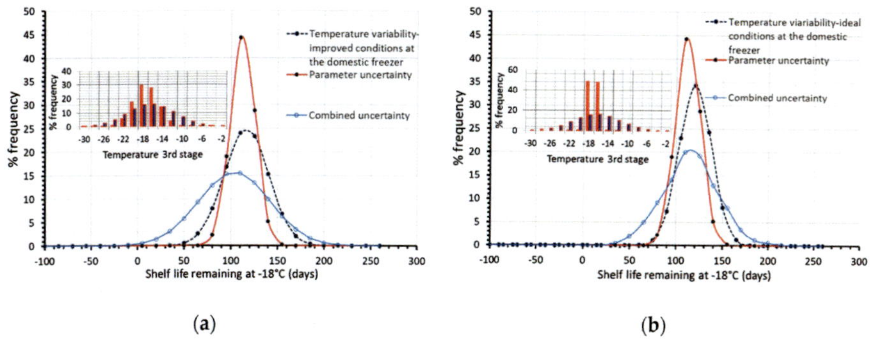

Figure 8.3: Establishing shelf life: Use of Predictive modeling.

(Source: Giannakourou, M., & Taoukis, P. 2020)

Predictive microbiology is premised on the idea that the reactions of populations of microorganisms to natural conditions are repeatable, and that by defining foods in contexts of those variables, it is possible to determine their responses in other comparable environments based on previous observations. This knowledge is encapsulated in mathematical models that allow for the prediction of microbial population behavior in foods over time (Kuijt, 2009). Predictive microbiology is an effective instrument for microbial food safety and quality assurance in and of itself, as well as in conjunction with threat analysis and critical control point initiatives, barrier technology, and quantifiable microbial risk assessment.

Microorganisms' responses to specific environmental variables like temperature, pH, and water activity are described in predictive microbiology. To visually characterize these responses, predictive microbiology employs statistical and mathematical (constructed with information from laboratory tests) and computer programs. Laboratory analysis, as well as the instruction and judgment of a seasoned food microbiologist, are not substitutes for predictive microbiological models. Predictive microbiological models should only be used by educated, qualified practitioners who are aware of the limitations of their application

A prediction must only be employed as a reference to the reaction of microorganism(s) to a specific array of external conditions throughout all predictive microbiology. Before using them, it is strongly advised that you consult with a competent body. To assess the safety of foods and/or data processing, food businesses must never depend entirely on any predictive microbiological model. Pathogen growth, survival, and inactivation in food necessitate the following:

- Determination of the product's intrinsic and extrinsic properties, factoring in storage and handling conditions, contamination risks, and anticipated shelf life (Kendall & Dimond, 2007).
- Review of published research and scientific data on the survival, growth, and neutralization of microorganisms of interest.

Food businesses should conduct additional studies if necessary based on these studies, which could include:

- Microbiological tasting and assessment in a laboratory
- Microbiological modeling that predicts outcomes
- Challenge tests to determine whether microorganisms of concern can grow or live in food under natural and probable distribution and storage circumstances.

Microbiological tests performed in laboratories are commonly used to make critical decisions about food quality and safe shelf life. Microorganism growth, survival, and inactivation in foods, on the other hand, are repeatable responses (Howey, 2015).

Microorganism growth, survival, and inactivation can be predicted using predictive microbiological models that quantitatively define the cumulative impact of particular environmental conditions. A predictive microbiological model in product design will help a food company evaluate the security and stability of additives and distinguish those that have a long shelf life.

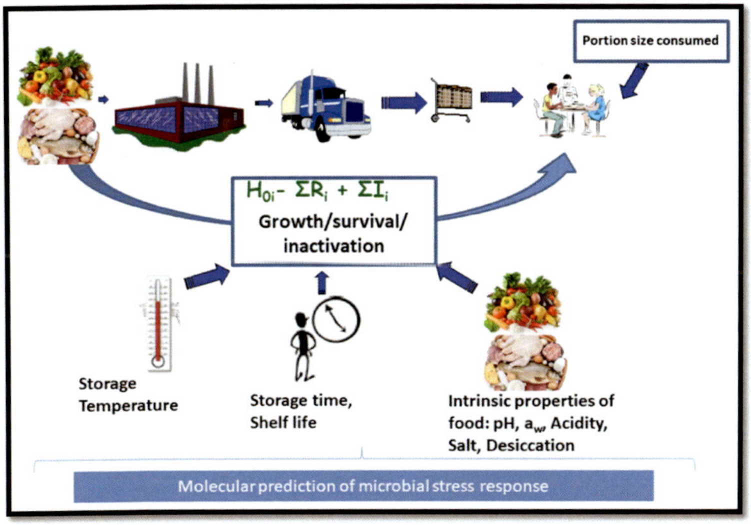

Figure 8.4: Factors affecting microbes in foods.

(Source: Njage PMK, Leekitcharoenphon P, et, al., 2020)

Whenever the shelf-life has been defined, but the product is then subjected to a slight process or formulation alteration, predictive microbiological models can be useful (both planned and unplanned with the absence of process control). After that, a predictive microbiological model can be used to see if the change has any impact on the product's safety and shelf life (Zhukov et al., 2018).

Product developers can use predictive microbiological models to identify the combinations of hurdles that can lead to preferred shelf life. If needed, these special parameters can be measured by experiment, lowering the cost of challenge studies.

8.3 DEVELOPMENT AND LIMITATIONS OF MODELS

In most cases, predictive microbiological models are built on the assumption that microbial responses are predictable. While predictive models can reduce the cost of microbiological testing when evaluating shelf life, the model's predictions may not always be precise due to microbial responses that are incongruent and differences in the growth media. According to research, this is one of the main reasons why certain predictive microbiological models perform poorly in correctly estimating pathogen survival, growth, or inactivation in food products (Weinbreck et al., 2010).

Moreover, before being used to aid in food safety decisions, predictive microbiological models should be validated. Model predictions are compared to experimental data that were not used in system design during validation.

8.3.1 Shelf Life

The term "shelf life" refers to the amount of time that a product can be consumed after it has been manufactured. The product must continue to be safe and have acceptable sensory, microbiological, and chemical properties.

Each product that can sustain the development of pathogenic microorganisms is required to have a "use-by" date after which it cannot be sold or consumed. This date is set by the manufacturer, who must be cognizant of all factors that may influence microorganism growth; understanding may be confined when creating a new product.

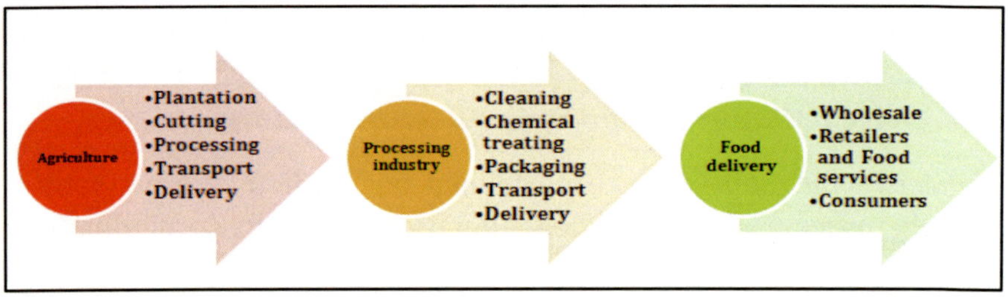

Figure 8.5: A supply chain for perishable food. (Source: Haji M, Kerbache L, *et, al.,* 2020)

The shelf life of a food product is among the most pertinent factors that determine its quality and, in some cases, its safety. The density of the major spoilage microorganisms has a big impact on shelf life. Mathematical models for bacterial activity are required elements for accurate shelf-life prediction and estimation in real supply chains. The mathematical models for describing microbial growth in food products have taken a lot of time and effort to develop (Villalobos-Delgado et al., 2019). For microbial growth, there are 2 kinds of mathematical models: deterministic and stochastic. In the particular instance of deterministic models, the result is a single growth curve or only one microbial concentration quantity at any given time.

Stochastic models, on the other hand, include a probability component in the analysis, with the result being the probability or probability density function (PDF) of having a specific microbial concentration at any given time.

The Baranyi and Roberts microbial growth model is a classical approach with four separate parameters: preliminary and maximal microorganism concentrations, lag phase, and rate of growth. The model must be calibrated for each trial or shipment because the lag phase is predicated on the physiology of the microbes at the start. Environmental conditions and the type of food have an impact on the growth of a particular microbial population (Wrona & Nerín, 2020).

Temperature is a critical environmental factor affecting microbial growth. Because the ambient temperatures along the real distribution chain are dynamic, it is critical to use models that can account for temperature variations to accurately predict shelf life. Microbial growth information under incessant environmental conditions is created to achieve the criteria in the main model. This procedure is carried out at various temperatures.

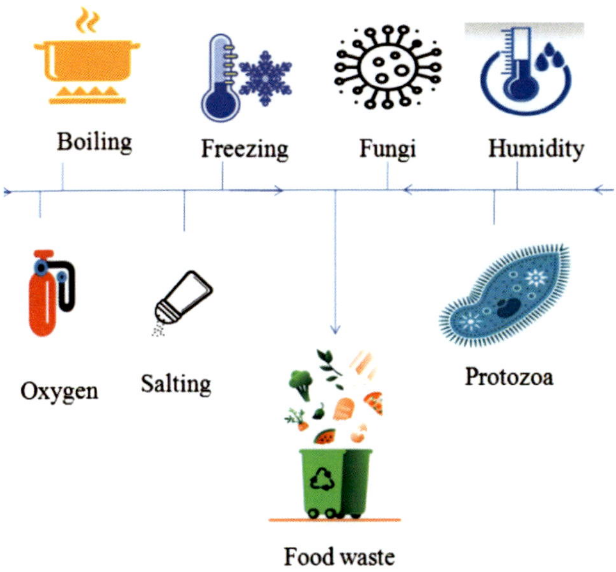

Figure 8.6: Food spoilage process. (Source: Sonwani E, Bansal U, *et, al.,* 2022)

The individual variables from the main models for various temperatures are procured by fitting the laboratory data collected in this manner on the primary model. The information for static temperatures is made to fit the square root model to determine the thermodynamic properties of the growth rate. Because the temperature profile in the distribution chain is not consistent and unforeseen temperature abuse could take place, using mathematical models for microbial growth to predict shelf life is of special significance.

The necessary temperature profile for shelf life evaluation is created by blending real-time supply chain temperatures up to the current time instant with a temperature profile gathered from prior supply chain monitoring of temperature from the present time to the end of the supply chain. The quality and safety of food products could be significantly improved as a result of this (Winterhalder et al., 2015).

Almost every consumer has come across spoiled food if it was moldy pastry, sour milk, bad fish, or the dreaded "leftovers" tucked away in the back of the refrigerator. A negative physiological or psychological reaction could have occurred if any of these were consumed. If pathogenic organisms were prevalent, a memorable bout of some unpleasant gastrointestinal ailment could have ensued. Foodborne pathogens can also be deadly in rare cases. Food spoilage, on the other hand, does not always indicate the presence of pathogens. Even so, realizing shelf life and stability can minimize the chance of coming into contact with spoiled or unsafe foods. The shelf life of food refers to how long it can be stored safely. Food develops characteristics at the end of its shelf life that is considered inappropriate or unappealing, such as alteration in taste, aroma, mouthfeel, or appearance. The change could be caused by microbiological, chemical, or physical factors. The above characteristics exemplify microbiological spoilage. Rancidity and freezer burn are instances of physical and chemical deterioration.

Many foods require identifying their microbiological shelf life at some point in time. The determination could be needed all through product development after they've been launched, or at any point in between. Product design goals, formulation changes, packaging or storage changes, changes in microbiological metrics, or a simple eagerness to learn are all factors for establishing shelf life. To reach a beneficial outcome, various factors must be recognized in the experimental setup of the shelf-life investigation, irrespective of when or why.

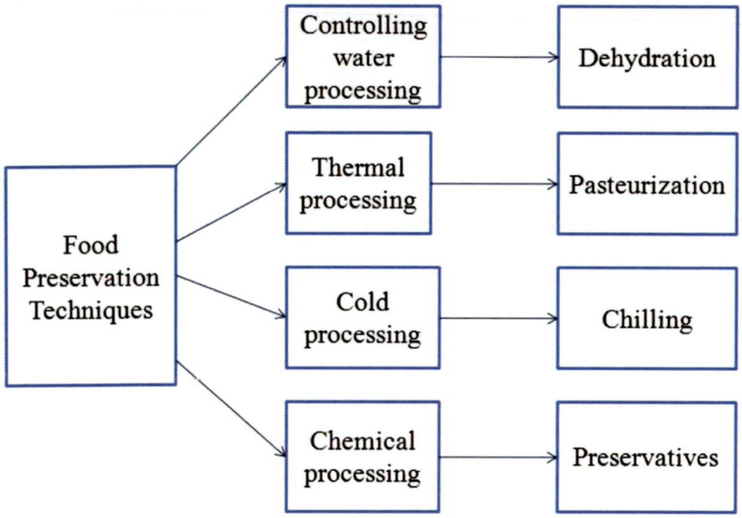

Figure 8.7: Food preservation techniques. (Source: Sonwani E, Bansal U, et, al., 2022.)

8.3.2 Spoilage Susceptibility

Temperature, moisture, acidity, nutrients, and time are all important factors in microorganism growth. Cultural circumstances have to be within a specific range for microorganisms to grow. The development will not occur if the minimum criteria are not met.

In general, organisms thrive at temperatures ranging from 0° to 55°C, pH levels of 2 to 10, and water activity levels greater than 0.6. These limit ranges aren't unconditional, and the boundaries that surround them aren't always well-defined. Optimized growth happens in the middle of the different ranges, and as the limits are approached, growth slows. In most foods, nutrients are not scarce, but inhibitory substances can prevent the food from being properly utilized (Wibowo et al., 2018).

Some organisms require oxygen to grow. Others think it's a waste of time, while others think it's poison. The ambiance or air trapped in the product can provide oxygen and other gases. Various groups of organisms are activated or neutralized, and their growth projections are speeded up or slowed, depending on the composition of the food, pH, water, and other factors.

The potential for spoilage of a food item can be determined by evaluating it in the context of its microbial growth requirements. Studies, on the other hand, are usually required to confirm expectations.

8.3.3 Qualifying Criteria

Consumers' ultimate criteria for judging a food's acceptability are its taste, odor, and appearance. Organoleptic assessment of a food item is a direct way to determine shelf life in the research lab as well. Food is prepared and inspected for changes in looks, aroma, texture, and taste till it becomes unacceptably bad (Uebele et al., 2013). Those who are acquainted with the preferred characteristics of the product can easily make the organoleptic determination. However, because tastes, expectations, and the ability to discern changes vary markedly, shelf life predicated on organoleptic assessment may differ wildly from customer to customer food's microflora — bacteria, yeast, and mold — develop and metabolize nutrient content, and the organoleptic value of the diet changes. The sensory alterations may be subtle initially, but they ultimately render the food unpalatable. Sensory changes are usually not noticeable till the microbial population has grown significantly. The multitude of organisms needed to spoil a food item differs based on the type(s) of microorganisms present. Microbial density can be used to estimate shelf life. 10 million bacteria per gram, 100,000 yeast per 1 g, or observable mold, as a rule of thumb, indicate the end of microbiological shelf life (Turner et al., 2020). At these levels, the product is likely to degrade noticeably. The organoleptic evaluation determines if the changes are acceptable.

In some foods, a high number of microorganisms is common, but in others, it indicates deterioration. As a result, even in the utter lack of offending organoleptic changes, it is necessary to ascertain the microbiological condition of food as it approaches the end of its shelf life. The shelf life of an item should be ascertained by sensory attributes and microbiological investigation to meet customer requirements of the highest quality.

8.4 DESIGNING A STUDY

When trying to design a microbiological shelf-life study, numerous considerations must be taken into account. Temperature, moisture content, time, kinds of microorganisms, the viability of analyses, sampling, and reproducibility are just a few of them. Because of the numerous variables that must be considered, shelf-life studies for every product must be specifically built for that product.

Temperature. Water, acidity, temperature, nutrients, preservatives, and atmosphere are all factors that influence microbial growth, but temperature becomes primarily remedied at the stage of product formulation, processing, and packaging. In most shelf-life studies, these variables aren't changed on purpose. The length of perishable foods' microbiological shelf life is usually determined by their storage temperature (Amani & Gadde, 2015).

In general, the rate of microbial growth increases as the temperature rises. Organisms develop very gradually or not at all at temperatures close to freezing. Metabolic action and growth rate hike as the temperature approaches the optimum. Growth is accelerated at this temperature. As the temperature rises above the ideal level, the rate of growth slows. Growth stops at a certain maximum temperature, and higher temperatures start to kill the cells. Each organism has a different minimum, optimum, and maximum temperature spectrum for growth. Furthermore, differences between isolates of the same species may be noted. The important point to remember about temperature and population increase is that when a product's storage temperature changes, not only does shelf life alter, but so do the kinds of spoilage flora.

The most useful shelf-life data is gained for products kept at their intended storage temperature because of the crucial connection between growth rate and storage temperature. Refrigerated products are kept in the refrigerator, while room temperature items are kept at room temperature. Small temperature changes in storage can have a big impact on shelf life. The difference between long shelf life and untimely spoilage may be determined by a few degrees.

Unfortunately, refrigerator and ambient temperatures are not streamlined in the real world. Refrigeration can range from 28 to 55 degrees Fahrenheit, and room temperatures can range from 60 to 95 degrees Fahrenheit. Even though both temperatures could depict refrigeration conditions, shelf life at 30°F could be very distinct from that at 50°F.

The temperatures used must be known to create a constructive study and to compare various studies. This is easiest to achieve if the study temperatures are constant and not variable. Coolers and room temperature storage temperatures are typically 40°F (4°C) and 75°F (24°C), respectively (Timakova et al., 2020).

Temperatures frequently cycle with both low and high during the distribution and retail display storage of perishable foods in real-world conditions. In laboratory shelf-life studies, temperature cycling introduces circumstances that make information assessment complicated, and it is usually not advised. When many storage temperatures are being used, much deeper comprehension of shelf life can be acquired. Studies on refrigerated foods can be carried out at three different temperatures (38° to 40°F, to 55°F). 75°, 85°, and 95°F are ideal room temperatures. In many cases, conclusions about other temperatures can be made based on the information gathered.

8.5 WATER

Many foods' characteristics are determined by their water content. Some foods are presumed to be dry, while others appear moist and contain water. Water is necessary for microbial growth, and the amount of free water in a food item can change its susceptibility to spoilage. A dry product that does not spoil, for example, will most likely spoil if it becomes damp. A moist food, on the other hand, will not spoil if it dries. Food packaging plays an important role in moisture control and has a big impact on shelf life. Between the atmosphere and the food, moisture is exchanged. This interaction keeps going until the food and the environment are in balance (Angourakis et al., 2015). The quantity of air in hermetically sealed packages is sparse, and the smaller the head space, the faster food and air reach equilibrium. Humidity control does not need to be considered as a study variable for hermetically sealed samples as long as the package stays intact.

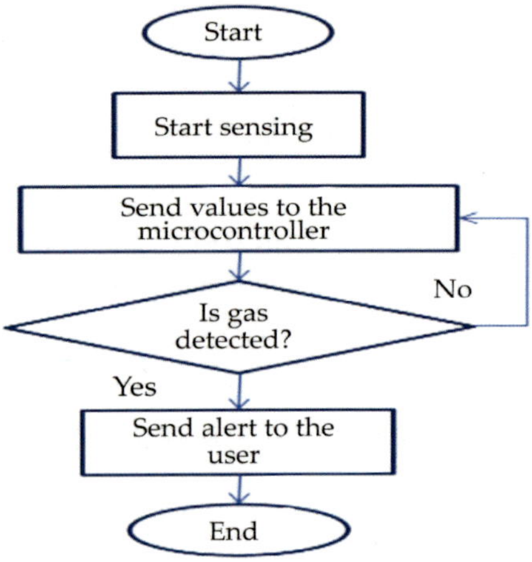

Figure 8.8: Humidity sensor. (Source: Sonwani E, Bansal U, *et, al.,* 2022)

Most foods are packaged to stop the rate of water interaction, resulting in minimal moisture exchange over the product's lifetime. Foods that are confined to temperature extremes, subjected to the surrounding air (such as cakes, pies, and pastries), or wrapped in air-permeable containers need humidity control and/or monitoring. A pragmatic variation for experimentation is 40 percent, 60 percent, and 80 percent relative humidity.

Duration. The duration of a study should at least equal the benchmark shelf life of the food under consideration. If a refrigerated item kept at 40°F is to have a 60-day shelf life, the study should be crafted for at least 60 days (Sutton et al., 2016).

Similarly, if the study is expected to last six months at room temperature, it should last at least that long. If demands were met and the point of spoilage must be determined, a study may be intended to surpass the shelf-life goal.

It's pointless to continue an analysis if a product fails midway through a planned shelf-life study. If the item is stable during one portion of a study (for instance, no microbial activity), the study should be progressed to the next portion. Sterile products don't need to be tested again after the time it takes for any contaminating microorganisms to outgrow. It's not uncommon for microbial levels to remain constant, or even decrease, for several hours, days, or weeks before starting to grow.

Because microbial growth is influenced by temperature, the microbiological shelf life of a food item intended to be kept at one temperature cannot be ascertained more rapidly by storing the food at a higher temperature. While it is correct that organisms grow more quickly in warmer environments, it is currently impossible to predict the outcome for a different incubation temperature. For a rise in temperature of 18° to 20°F, a two- to four-fold rise in growth rate is projected.

If the shelf life is 10 days at 60°F, it is approximated to be 20 to 40 days at 40°F. This wide range of possibilities for a prediction isn't very useful (Calligaris et al., 2016).

Furthermore, it is completely possible that the organisms which started growing at the elevated temperature (60°F) do not grow at the preferred storage temperature (40°F), or that the microbes that did grow at the higher temperatures (60°F) are not the primary spoilage organism. Expedited microbiological shelf-life projections are therefore useless.

8.6 METHODICAL APPROACH

Agar plating procedures are most commonly used to monitor microbial growth in foods for shelf life estimation. For the number of viable organisms existing at the time of analysis, the processes are quantitative. Because the growth requirements of different types of microorganisms found in food differ, there is no single procedure for enumerating all microorganisms. The aerobic plate count, which recognizes organisms that establish colonies on plate count agar incubated at 35°C for 48 hours, is a simple and useful procedure.

Many organisms are undetectable when the aerobic plate count is done in a mesophilic environment. Organisms that only grow at low or high temperatures, most lactic acid bacteria, strict anaerobes, yeast, and mold are among them. As a result, plating procedures are typically chosen based on the kinds or types of organisms that are expected or known to be existent in the food. It is very possible to have noticeable microbiological spoilage but no experimental evidence to back the organoleptic findings if the "right" procedures are not chosen.

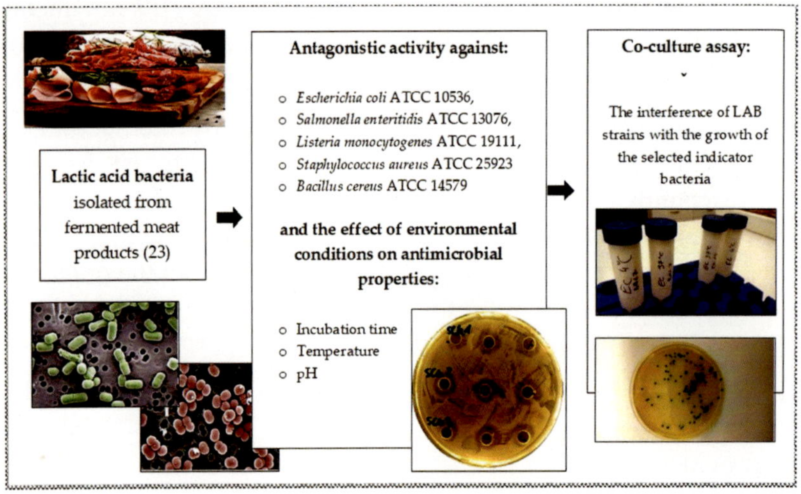

Figure 8.9: Plating assays . (Source: Łepecka A, Szymański P, Rutkowska S, et, al., 2021)

8.7 SENSIBLE SAMPLING

To detect significant microbiological events, the frequency with which a food item is examined for microorganisms in the shelf-life study must be carefully chosen. The general growth cycle of a group of bacteria should be understood to better explain why this point is important. There are four stages to the growth cycle. The "lag phase" occurs at the start of the cycle. During this time, no increases in cell numbers have been observed. The count rises dramatically in the second phase: one cell will become two, then four, and so on. Although neither rate of growth nor the number of cells continue to increase during the stationary phase. Growth usually stops at a density of 1,010 bacterial cells per gram or 106 yeast cells per gram (Castrica et al., 2021). Because cell viability reduces in the last phase of the cycle, it is appropriately dubbed the "death phase." The cells will not start a new cycle until they are diluted into a new growth medium, like food.

The food would seem to be microbiologically stable while the microorganisms in it are in the latent period. The item starts to change and is deemed unstable once the cells enter the growth period and start to multiply. The food will generally spoil at some point along the microbial curve. As a result, the duration of the lag phase, the growth rate, and the microbial count at the end of the growth phase is the most important aspects of the microbial growth cycle for shelf life. The end of shelf life, which is defined as 10 million bacteria or 100,000 yeast per gram, typically occurs near the end of the growth phase.

The food is sampled regularly to quantify the number of organisms present in order to identify the distinct transition points along the growth path. If you wait too long between samplings, you risk underestimating or overestimating shelf life. The accuracy of the shelf-life determination will improve as more analyses are completed. For sample gathering and processing, the expected shelf-life time is split into five to twelve intervals for most foods. The quantity of intervals selected is usually a guess based on previous interaction with similar foods.

Microorganism dispersion in a sample of food, or even between samples of food from the very same production batch, is not always consistent. If one out of every five containers of food solution includes a spoilage organism, only one out of every five bottles will show signs of spoilage. Another example is a spoilage organism current in a solid or viscous food that causes localized spoilage while leaving the rest of the food unspoiled. The distribution of microbes within the lot must be considered when designing sampling plans. When the initial levels are less than 10 cells per gram, that is the normal sensitivity of the agar plating processes used for analysis, this is especially important. The most distinguishable analytical unit is single packages (Chang et al., 2022).

If the product has an even dispersion of organisms, a meaningful study can be conducted with fewer samples. Each analytical unit in a homogeneous product is much more likely to be exactly equivalent to the next. Liquids, for example, are homogeneous if they are blended each time before sampling. The content of the whole will be represented by each unit of liquid. An unmixed sample, on the other hand, may have an inconsistent dispersion from top to bottom or side to side. It's possible that thorough mixing won't be possible in a viscous sample, likely to result in a non-dispersion of organisms. A single sample could not be representative of the entire product lot. To represent the whole, multiple subsamples must be analyzed for these products. To achieve an appropriate depiction of microbiological activity in a heterogeneous product, at least 3 cautiously selected samples are required.

A shelf-life test performed on a single batch of food is legitimate for that batch and any other identical production lot. The shelf-life duration could differ if the type or number of microorganisms differs significantly between batches. The precision of the prediction is always improved when the study is replicated. Periodic shelf-life determinations help to ensure that the product stays consistent in terms of spoilage rate over time. Changes in a product's formulation, processing, or packaging conditions necessitate reevaluating its shelf life.

8.8 MAKING A CHALLENGE

Even if the temperature, pH, water, and nutrient levels allow for microbial growth, a food item can have an extremely long shelf life. The long shelf life could be due to the lack of microorganisms in the tested samples, or it could be since contaminating organisms will not develop in the specific product formulation. A microbiological shelf-life study, where a product is confronted by inoculating it with suitable spoilage organisms, is required to understand the reliability of these foods in the occurrence of possible contamination (Dawson et al., 2018).

The item is inoculated with known spoilage microorganisms in a challenge study. After that, the inoculated samples are treated and stored according to the shelf-life study's recommendations. The addition of organisms to the foods increases the number of variables in the study. If the product has an even dispersion of organisms, a meaningful study can be conducted with fewer samples. Each analytical unit in a homogeneous product is much more likely to be exactly equivalent to the next. Liquids, for example, are homogeneous if they are blended every time before sampling.

The substance of the whole will be represented by each unit of liquid. An unmixed sample, on the other hand, may have an inconsistent dispersion from top to bottom or side to side. It's possible that thorough mixing won't be possible in a viscous sample, likely to result in a non-dispersion of organisms. A single sample could not be representative of the entire product lot. To represent the whole, multiple subsamples must be analyzed for these products. To achieve an appropriate depiction of microbiological activity in a heterogeneous product, at least 3 cautiously selected samples are required.

The number of microorganisms introduced to the food is typically much higher than what would ordinarily be encountered in processing due to contamination. The inoculation levels used are typically greater than 10 per gram, allowing for easy detection of the struggle organisms; 10 organisms per gram denotes the sensitivity limit of the commonly used agar plate count procedures. Lower levels can be identified, but at a much higher cost and with a lower degree of accuracy (Sewald & DeVries, 2003).

The product formulation is resistant to microbial growth if the quality of the challenge organisms does not rise during shelf-life storage.

In the manner that the number of microorganisms does not increase, it is stable. However, if there are enough organisms, it is still possible that the non-growing cells' metabolic activity will cause unpleasant changes in the product.

The organisms that cause spoilage in most foods do not start multiplying right away. Instead, before growth is noticed, the count continues to remain moderately constant for some time. The lag stage of development of the microbial growth cycle is analogous to the period of no growth. Some of the challenging organisms may die just when added to the test sample.

If the test has a low starting inoculation level and die-off takes place, it's easy to assume the product is stable. This error can be avoided by using high inoculation levels. Even if a 100-fold die-off is observed, a level of about 10,000 cells per gram is beneficial for observing either reductions or increases in levels.

The shock-induced by a sudden shift in the environment for which the cells have not been preconditioned is most likely the cause of death after inoculation. Adapting the organisms to the product's nutrients, acidity, or water activity can sometimes prevent or reduce the die-off. In the real world, both unadapted and adapted organisms can contaminate products (Sumner & Jenson, 2011). The use of organisms that are not particularly adapted for product growth resembles organisms that originate in the environment and enter the food via contact. Adaptation simulates contamination from one product to another

Pathogen challenge studies are similar to studies using spoilage microorganisms in that they are used to measure the actions of those microorganisms in foods and formats.

The goal of using pathogens in food is to track their growth, inhibition, or death. Salmonella, *Listeria monocytogenes*, *Staphylococcus aureus*, *Bacillus cereus*, *Clostridium perfringens*, *Yersinia enterocolitica*, *Clostridium botulinum*, and *Escherichia coli* are some of the most commonly used pathogens. If the pathogens will not develop, the food is regarded as stable in terms of the food system's ability to prevent them from growing.

The length of the pathogen's lag phase is influenced by preventing changes in the organism's environment. The longer the lag phase, the greater the inhibition. Even temperature differences between the incubation temperature used to propagate the organism for the research and the product's storage temperature can affect the length of the lag period. As a result, the preparation conditions for the organisms should be carefully selected.

8.9 MATHEMATICAL MODELING

It is simple to measure the effects of the environment, temperature, water, pH, preservatives, and nutrients on microbial growth. After gathering enough data, a mathematical equation for growth can be derived. Plugging the required food variables into the equation could allow for a quick prediction of shelf life. There are mathematical models for bacterial pathogens that provide some of the growth-controlling variables (Sarkar et al., 2022). The models are intriguing, but they aren't always accurate when it comes to food. They're useful for "what if" games like "How much longer does lowering the pH of the food from 5.6 to 5.4 delay the growth of Salmonella?" Models will undoubtedly become more useful as time goes on. However, it will be a long time before they can completely replace experimental shelf-life studies.

A carefully planned and implemented set of experiments studies is required to accurately predict shelf life. If changes in ingredient, formulation, processing, packaging or storage are expected, the shelf life should be reevaluated. The information gained from these studies ensures that the product delivered to customers is both safe and of high quality.

CHAPTER 9

HOW TO ENHANCE THE SHELF-LIFE OF FOOD

CONTENTS

A food's shelf life can be defined as the amount of time it takes for its quality to deteriorate to the point where it is no longer fit for human consumption. To identify undesirable characteristics, objective indicators such as off-odors and off-flavors, variations in texture attributes, and visual alteration are used. Microbial, chemical, or physical interactions are to blame for all of these changes. Temperature, pH, aw, microbial contamination of raw materials, and cross-contamination during manufacturing, storage, and marketing are just a few of the many variables that can impact a product's shelf life (Roccato et al., 2017). They include: The economics (production costs and waste management), as well as the health risks associated with food consumption, are both significantly impacted when shelf life is reduced.

9.1 MEAT AND SEAFOOD STORAGE AND SAFETY

Meat and seafood are particularly prone to spoilage because of their chemical composition and high water content. Microbial metabolism and oxidative processes affecting the lipid components are the primary sources of degradation in these products. When it comes to preserving the shelf life of fresh food, a proper cold chain is the most important factor in ensuring that spoilage doesn't begin until it's too late (Pratiwi et al., 2015).

Figure 9.1: Meat and seafood.

Vacuum or modified atmosphere (MAP) packaging typically enhances the cooling effect during commercialization. Although CO2 is typically used to provide the antibacterial effect, other gas compositions may be used in the latter situation. Gram-negative aerobic bacteria can be inhibited or hampered in their growth by atmospheric carbon dioxide concentrations greater than 20 percent, and this can alter the dynamics of other microbial communities. Oxygen levels up to 60 percent are needed for certain species of meat to maintain their color. However, the product's oxidation, which is related to the meat content and the time it was preserved, should be examined.

Extending the shelf life of processed foods necessitates the use of temperature control techniques like chilling/freezing, pasteurization, and sterilization. There are a number of potential solutions to the aw reduction problem, including the use of non-thermal treatments (e.g. HHP) and a variety of methods for controlling pH and redox potential, as well as the use of preservatives and MAP (i.e., HHP, ionizing radiations) (De Corato, 2020).

Fresh meat and seafood are being studied for ways to keep them fresher longer due to consumer and market demand. In addition to the desire for a "clean label" and a desire for healthy and "natural" procedures (reduced technical effect in the concept of minimally processed foods and reduction/elimination of preservatives), the primary drivers are economic considerations (reduction of waste).

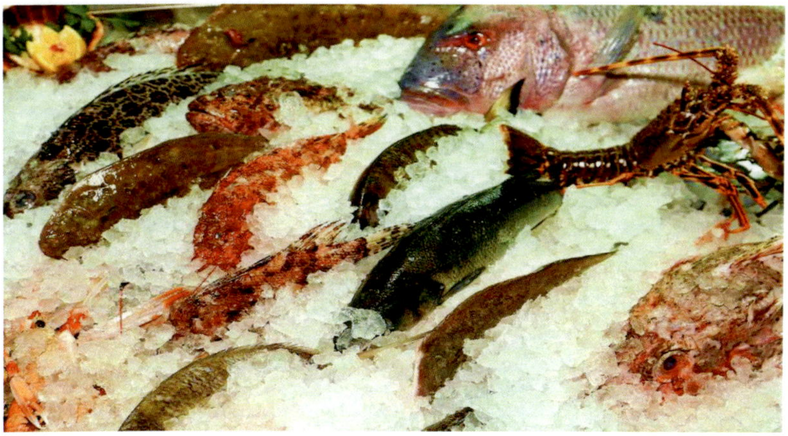

Figure 9.2: Preservation of fish.

Bioprotective microorganisms are gaining significant traction in the food industry as a method of extending the shelf life of meat and seafood. By including beneficial bacteria (usually lactic acid bacteria) in a product, manufacturers can keep spoilage-causing microorganisms (such as pathogenic bacteria) from taking hold and spreading. Because of their ability to compete in the specific food matrix and their ability to produce antimicrobial substances (such as bacteriocins, organic acids, and volatile compounds), the cultures can achieve this goal. It is also important to select bioprotective cultures to minimize the impact on the product's overall organoleptic characteristics (Peleg et al., 2017). Compounds derived from plants and other natural sources can also be used to extend the shelf life of meals. Some of these molecules may be antibacterial, while others may be antioxidative.

There are a lot of terpenoids, terpenes, and phenylpropanoids in essential oils in this state. Despite their well-documented antibacterial properties, little is known about their mechanisms of action and interactions with other stabilizers. To expand and standardize their use in industry, more research is needed. Natural phenolic compounds, which include at least one or more aromatic rings with hydroxyl substituents and are produced by plants in their secondary metabolism, are responsible for the antioxidant activity of many of these chemicals. These chemicals can be used in a variety of ways, which is interesting. Although they can be used as an ingredient, they can also be used during cooking, such as marinades. Instead, they can be incorporated into active packaging or food matrix after being encased in nanoparticles (Omotilewa et al., 2018). Organoleptic properties and interactions with proteins and lipids, which may significantly reduce their antibacterial efficacy, are the main concerns about their use in meat and fish. Chitosan is a different method of preserving food quality and shelf life. Fungi, mollusks, and crustaceans all contain this chitin-based polymer.

It is GRAS-certified by the USDA and has antibacterial properties against bacteria (both Gram-positive and Gram-negative) (both Gram-positive and Gram-negative). As a food additive, it is used in a wide range of applications. Meat and seafood can be contaminated with a variety of pathogenic and/or toxin-producing microorganisms, which raises significant safety concerns. *Escherichia coli* O157:H7 and *Campylobacter jejuni* are the Gram-negative bacteria responsible for the majority of severe illness outbreaks in Europe. One of the most worrisome Gram-positive bacteria is *Listeria monocytogenes* and the most common foodborne toxins are *Staphylococcus aureus* and *Clostridium perfringens*. The presence and development potential of these species must be strictly regulated if food safety is to be ensured in the food industry.

Food shelf stability and perishability are influenced by extrinsic factors such as gaseous atmosphere, storage temperature, and relative humidity, as well as innate food properties. Shelf-stable foods have lower water activity and/or pH values when compared to perishable commodities. It has an unpleasant taste and odor, as well as gas accumulation and exudate discharge, as well as physical changes in shape, consistency, and color.

As a result of floral development or other chemical changes, stale meat, and fish release gaseous amines such as ammonia. Histamine, putrescine, and tyramine, to name a few, have been linked to hypertension and migraine. The purine breakdown product xanthine is used to assess the shelf life of meat and fish. To create a xanthine sensor, the enzyme xanthine oxidase was combined with a multi-walled carbon nanotube scaffold dispersed with chitosan–ZnO NPs (O'Brien & Bentley, 2015). By embedding ZnO nanoparticles in carbon nanotubes, electron transport between biomolecule active sites can be increased and accelerated (xanthine oxidase and electrode with good chemical stability). Chitosan-enabled bioadhesive sheets were used to successfully immobilize xanthine oxidase.

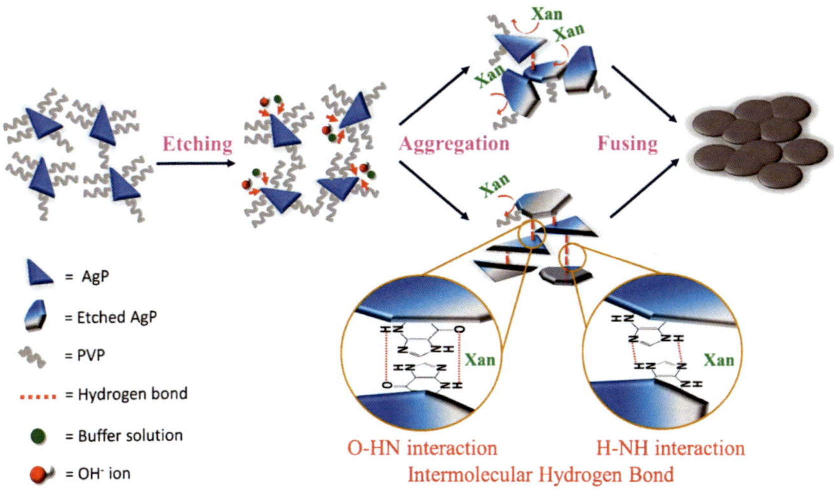

Figure 9.3: A xanthine sensor. (Source: Hsu H-C, Liao P-W, *et, al.,* 2020)

The detection limit of the sensor is 0.1 mm in the absence of matrix interference. When conjugated to AgNPs, xanthine oxidase was more stable at room temperature for 60 days.

The majority of food degradation is caused by high levels of moisture and gases, primarily oxygen. The detection of ethanol and oxygen was investigated utilizing ZnO- and TiO2-nanoparticles. It is possible to detect even the smallest amounts of oxygen retained in packaging or leaking of a product by combining TiO2- or SnO2-NPs with the redox-active methylene blue dye. Volatile organic gases such as ammonia, hydrogen sulfide, nitrogen dioxide, CO2, moisture, and formaldehyde, as well as ammonia and hydrogen sulfide vapors, have been detected using polymers (nylon-6, polyacrylic acid, polyvinyl alcohol) and ceramics (Ding et al., 2010). Changes in the oxidation-reduction potential of oxide semiconductors caused by gas absorption affect electric conductance. Polyacrylic acid nanofibers have an ammonia detection sensitivity of 130 ppb when measured with a QCM (Quartz crystal microbalance) (Ding et al., 2005). Fluorophores embedded in Perylene-based nanofibrils can detect gaseous amines at parts per trillion concentrations. Dampness has an effect on carbon-coated CuNPs in a tenside (detergent) film. Moisture causes the polymer to expand, causing the NPs to separate, resulting in a change in color absorption and reflection. The sensors can be used again once the moisture has evaporated.

9.2 PRESERVATION PROCESSES

Food preservation techniques can help to extend the shelf life of food. The first steps in the preservation process are agricultural practices and hygienic requirements for raw materials, as well as proper manufacturing techniques in the industry.

Heat, biocides, and modern conservation technologies are among the preservation methods used by the company to eradicate microorganisms. Low temperatures, low water activity, decreased pH, natural or synthetic preservatives, and changing atmospheres are examples of preservation methods that rely on limiting germ development. In some cases, a combination of multiple preservation methods is used to improve the efficiency of the preservation strategy (Nicoli, 2012).

Temperature, water activity, pH reduction, and preservative addition can be divided into two categories: traditional preservation procedures and so-called new preservation methods. Irradiation, pulsed energy emission, high pressure, innovative methods of applying heat and/or packaging, as well as various antimicrobials and nanotechnology applications, are among the most common methods. Hurdle technologies, on the other hand, combine different preservation techniques to optimize inactivation, reduce the treatment intensity of each operation, and thus limit the process's impact on food quality.

Traditional methods, for example, are cost-effective, safe, and well-established; they can be used for a wide variety of heat treatment applications. Nonetheless, there are concerns about the therapy's inconsistency and the resulting changes in chemical, physical, and dietary properties. According to recent market research, consumers prefer food that is fresh, nutritious, safe, and simple to prepare. Given the limitations of current technologies, particularly heat treatments, and customer expectations, the future of food preservation is contingent on evolving toward 'processing on-demand,' which means treating each product uniquely. Traditional preservation methods will continue to be the best option for some foods, but niche markets have emerged for developing technologies that produce food that is microbiologically safe, healthful, and retains the fresh product's qualities when applied to specific goods. These are the meals that are most likely to meet the consumer's requirements.

The process of preserving food begins with drying. Lowering the moisture content and water activity of dietary components can improve their microbiological stability. Microbial growth necessitates a water activity (aw) of about 0.6. Dehydration, on the other hand, can alter the texture, flavor, and other physical properties of food products. Foods with a water activity of 0.6-0.84 and a moisture content ranging from 15% to 40% can be consumed without rehydrating. IMFs are shelf-stable meals. Osmotic dehydration has several advantages over traditional dehydration, including lower processing temperatures, faster drying times, and a 20% to 30% reduction in energy consumption (Nirupama et al., 2010). In candied intermediate-moisture fruits, sugar is the primary humectant, whereas salt is used in intermediate-moisture vegetables and seafood. Without the use of artificial colors, flavors, or gelling agents, most fresh fruits can be transformed into IMF or candied fruit (also known as crystallized fruit).

The shelf life of a product is affected by its material, processing method, packaging, and storage conditions. Understanding and optimizing the relationship between these four characteristics is critical for achieving the required shelf life. Chemical, microbiological, sensory features, and nutrient losses must all be considered when determining a product's shelf life. Sensory analysis with a trained taste panel or chemical analysis (for example, measuring the synthesis of a specific molecule such as trimethylamine in fish) can both detect sensory quality changes.

9.3 PROCESSING METHODS

A variety of processing methods can be used to extend the shelf life of a food product. The most common methods are sterilization and fermentation. Pasteurization is a scientific method of destroying any germs that may be present in food by heating it to temperatures below 100 degrees Celsius.

9.3.1 Blanching

It's a common food processing step that involves placing an item in boiling water and applying a moderate amount of heat to it. This strategy's primary goal is to render enzymes inactive. The most common examples are vegetables and fruits. This is a novel way to present fruit. Canning, sugaring, drying, curing, pickling, and vacuum packing are all traditional methods of food preservation (Derkach et al., 2020). When it comes to food, you must choose a strategy based on what you have on hand.

Figure 9.4: Difference between blanched and unblanched.

9.3.2 Temperature

The most important factor in determining a product's shelf life is temperature. The temperature is adjusted at each stage of food processing and production. The most common method for extending the shelf life of perishables is to produce food in a temperature-controlled environment. Temperature abuse during the manufacturing process can significantly reduce a food product's shelf life. To ensure precise results, temperature control is used throughout the entire food preparation process.

9.3.3 Chemicals

Food chemicals are frequently used to extend the shelf life of food products. To ensure that chemicals do not degrade the quality of the object, they are used in conjunction with other processes. Many food manufacturers rely on antioxidants and oxygen absorbers to keep their products fresh and extend the shelf life of their products (El-Anany et al., 2009). Food manufacturers frequently include oxygen absorbers in small pouches with food packets to extend the shelf life of their products.

9.4 THE OPERATING ENVIRONMENT FOR THE PRODUCT HAS CHANGED

The packaging of a product determines how long it will remain fresh on the shelves. Fresh food manufacturers frequently pack their goods in an oxygen-free environment that they control. They use “Modified Atmosphere packing” to keep oxygen out of their finished product packaging.

To keep food fresh for a longer period in packaging, companies use a mixture of gases other than oxygen. Some businesses use vacuum packaging to keep their products fresh, such as fruits and vegetables, while also improving their appearance and attracting customers.

9.4.1 Product Irradiation

Food irradiation is the process of exposing food and its packaging to ionizing radiation. The practice of exposing food to radiation successfully removes germs or dangerous microorganisms that cause deterioration or potentially cause sickness after consumption. While the ionizing radiation never comes into direct contact with the food, it is nevertheless able to eradicate the living pathogens, insect infestation, and other contaminants that may be present (Nagi et al., 2012). This method improves the safety aspects of food commodities and serves an important role in extending the items’ shelf life. Irradiation of products is a common method for pasteurizing milk, canned fruits, and vegetables.

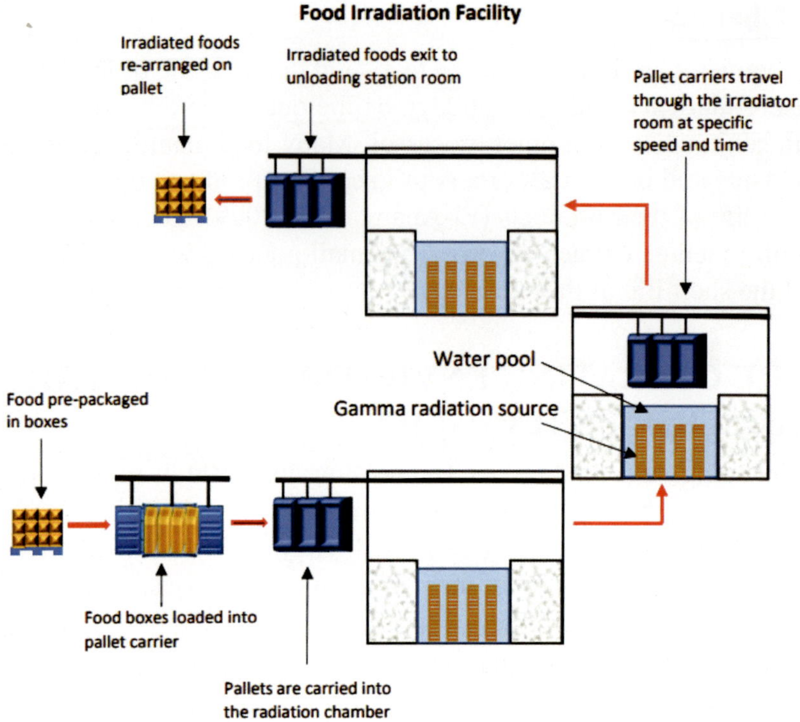

Figure 9.5: Food irradiation process. (Source: Balakrishnan N, Yusop SM, *et, al.,* 2022)

9.5 HOW IS THE SHELF LIFE OF FOOD PRODUCTS DETERMINED?

When new items are introduced to the market for the first time, shelf life is evaluated. Various spoilage organisms are examined in prepared meals for an extended period until their levels exceed safe limits. That's how it's done. Food processors use a variety of strategies to increase the shelf life of their products. Consider the following examples as illustrations:

- **FERMENTATION**- Microorganisms change the quality of a food product, preserve it by creating acids and alkalis, and add delicate tastes and perfumes to raw ingredients to increase its quality and value through fermentation (Fan et al., 2009). To guarantee good fermentation, temperature, pH, oxygen consumption, and carbon dioxide generation are all continuously monitored and managed.

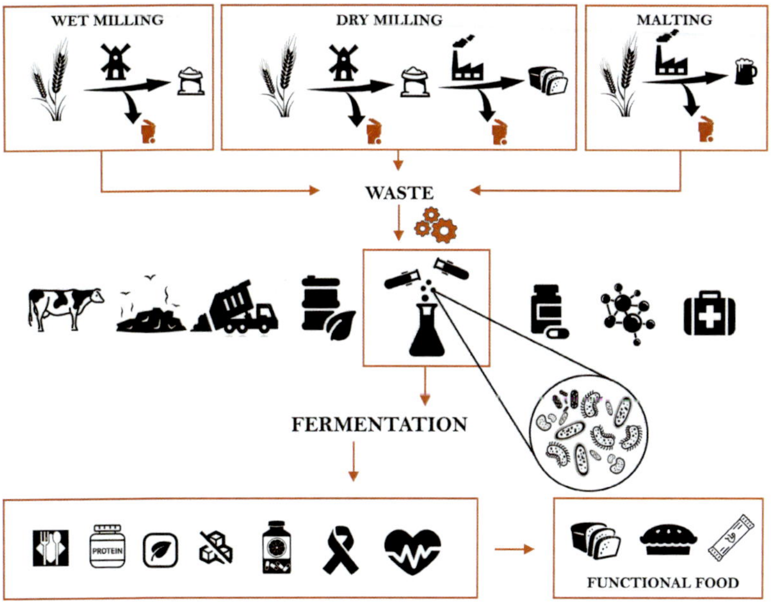

Figure 9.6: Fermentation of fruits. (Source: Verni M, Rizzello CG and Coda R .2019)

9.6 PROCESSING BY APPLICATION OF HEAT

Drying- It is a unit operation that reduces the moisture content of a food product to some amount by applying heat under regulated conditions to avoid the formation of spoiled organisms and some undesired enzymatic activity. The primary purpose is to extend shelf life by reducing water activity. Dryers of many varieties are used for this purpose, including freeze dryers, vacuum dryers, heat pump dryers, spray dryers, and drum dryers.

Pasteurization- A mild heat treatment at temperatures below 100 degrees Celsius is indicated to minimize the microbial burden in food. The flavor and nutritional content of the meal have only been minimally affected. Temperature and time decisions are set with both commodity value and microbiological risk in mind. You can use HTLT (High temperature, long time) to raise the temperature to 72°C for 15 seconds, and LTLT to lower it to 32°C for at least 30 minutes (Maertens et al., 2012). These are the most frequently used comb

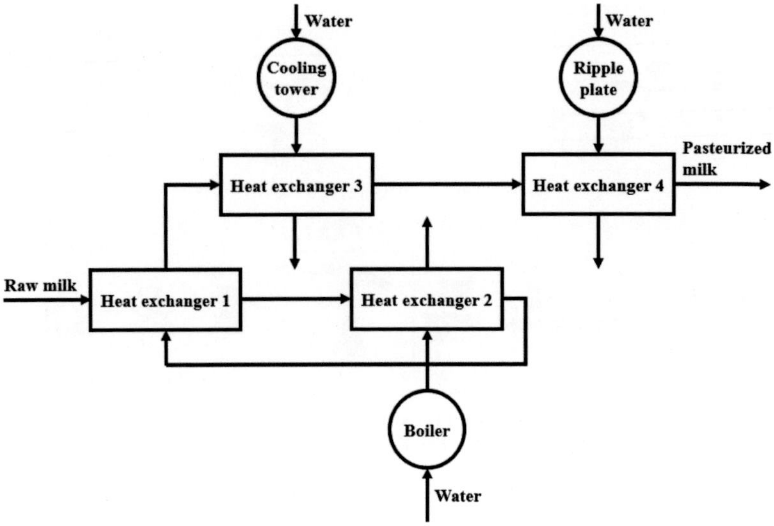

Figure 9.7: A Pasteurization scheme. (Source: Rasmussen EDJ, Errico M, et, al., 2020)

Blanching-In simple heat treatment, enzymes can be inactivated by using boiling water (pre-heat treatment). The most common food items prepared in this manner are vegetables. When fruits are sold, they are frequently unblanched. The amount of time it takes to blanch a vegetable or fruit is affected by many factors, including

- The type of vegetable or fruit. The particle size of the food
- What temperature should you use to blanch your food?
- The heating method

Blanching can also be referred to as precooking, parboiling, and scaling.

Heat sterilization- Food is cooked at a high temperature for an extended length of time to eradicate microbial and enzyme activity. As a result, sterilized food has a six-month shelf life. They are classified into two types:

First, there was dry heat sterilization, which used heated air with only a trace of water vapor.

A moist heat sterilizing agent is hot air that has been saturated with water vapor. Moisture is essential throughout the sterilizing procedure (Manzocco, 2016).

Dielectric, infrared and ohmic heating- Meal preparation can be done using either direct or indirect heating. You can use dielectric and ohmic heating to directly heat food (for example). If you wish to indirectly heat food, you can utilize radiation heating (for example) (example-infrared heating). These three strategies are used to transfer heat in many ways. When dielectric heating is applied, molecular friction develops between the commodity's water molecules, resulting in heat creation. Electrical resistance in food creates ohmic heating, which results in heat production. As previously stated, heat is created outside and then transmitted to the food product.

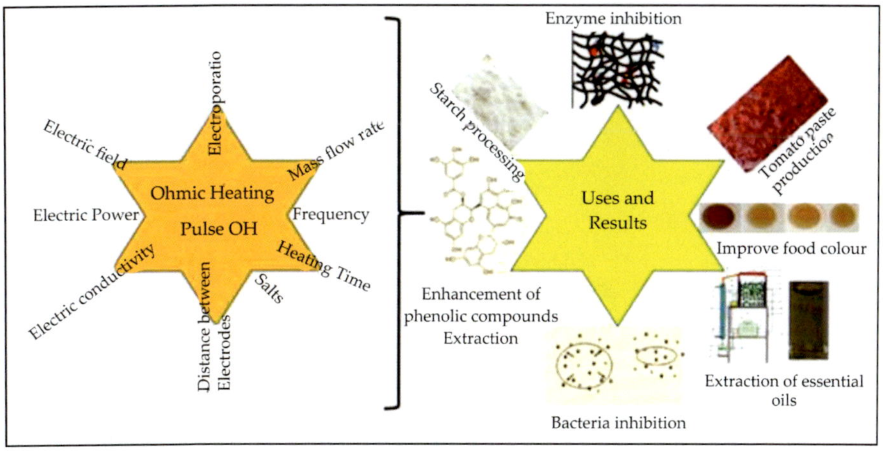

Figure 9.8: Ohmic heating. (Source: Alkanan ZT, Altemimi AB, *et, al.*, 2021)

9.7 PROCESSING BY REMOVAL OF HEAT

Chilling- When a meal is chilled, the temperature is dropped to between 1 and 8 degrees Celsius. This project's main objective is to reduce the rate of biochemical and microbial evolution. Other unit activities, similar to fermentation and pasteurization, are frequently employed in combination with it. Some of the advantages of chilling include: Flavor, color, texture, and form are all largely intact, as does the nutritional content. Keeping fresh goods for a long period is doable if they are preserved properly.

Freezing- When a food product is heated below its freezing point, a fraction of the water in it changes state and becomes crystalline, generating ice crystals. The ice-cold water is unfriendly to microbes and chemical reactions. When it comes to nutrition and flavor, conventionally kept meals can fall short of frozen alternatives. Frozen dishes frequently have a longer shelf life. The freezer is normally set to a temperature of -18 to -25 degrees Celsius (Manzocco et al., 2010).

9.8 POST-PROCESSING OPERATIONS

Waxing- Waxing is the process of applying wax to a fruit or vegetable to keep its moisture content. Fruits and vegetables seem more appetizing after waxing. Waxes are available in some formats. Beeswax (yellow and white) and carnauba wax (yellow and white) are also commonly used (Brazil wax and palm wax)

Wax derived from paraffin

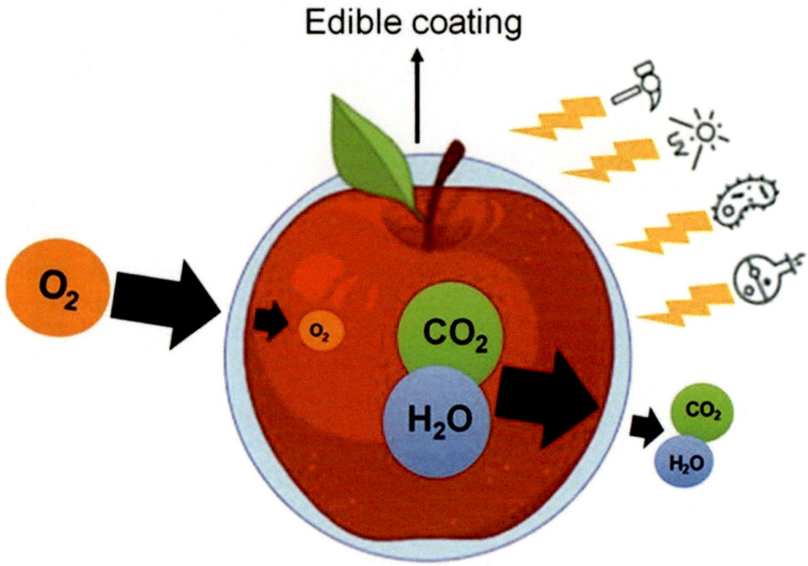

Figure 9.9: Paraffin wax(Source: de Oliveira Filho JG, Miranda M. *et, al.,* 2021)

9.9 SHELF-LIFE OF BAKED PRODUCTS

To extend the shelf life of baked goods, manufacturers, product developers, and process and packaging technologists must create appealing and tasty items that remain fresh for an extended period without acquiring any germs. A variety of factors influence the long-term shelf life of baked goods (Martins et al., 2008). Although some chemical reactions and physical changes are unavoidable, there are a few methods that can be used to postpone them and maintain a high-quality product life during the development of a new food product.

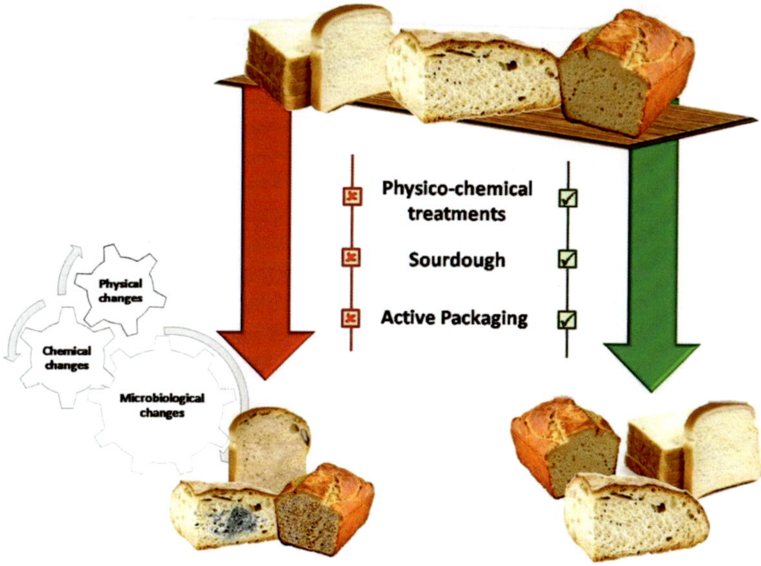

Figure 9.10: The shelf life of baked products. (Source:Melini V, Melini F. ,*et, al.,* 2018)

9.9.1 Staling

When most baked goods, such as bread and rolls, are exposed to room temperature, they become wet and spongy, a condition known as staling. In general, the higher the moisture content of the fresh product, the more pronounced the staling-induced changes. Because of their higher water content, bread, sweets, and cakes, for example, are more prone to spoilage than biscuits and crackers. Bread's crumb and crust age at different rates due to the same underlying cause (Lindström et al., 2006). Crumb staling is indicated by several physiochemical changes, including altered flavor and fragrance, increased hardness, and increased starch crystallization and retrogradation. To combat staling, packaging methods can be used to limit moisture loss and keep the crumb soft and pleasant. You can delay Staling while developing a bread product by keeping storage and transportation temperatures below 32°C.

Figure 9.11: Staled bread.

9.9.2 Emulsifier

Emulsifiers aid in the retention of moisture, which slows the aging process. Surfactants can be used to improve shelf life in a variety of ways, including increasing loaf volume and achieving an optimal gluten structure. The resulting crumb is softer and retains more moisture as a result of this method.

Figure 9.12: Food emulsifiers.

Gluten proteins obstruct protein-protein interactions due to ionic repulsion. Salt strengthens the dough by reducing ionic repulsion. Surfactants (DATEM, SMG) are fatty acids that bind to the hydrophobic regions of gluten proteins. Saturated fatty acids are preferred because unsaturated monoglycerides do not have the desired effect, such as increased air uptake. Oil-based cakes have a more delicate texture and a longer shelf life. This, however, harms foaming. Monoglycerides that have been acetylated or lactylate, for example, are excellent emulsifiers that can increase the oil's usefulness in baked goods (Liu et al., 2018).

9.9.3 Catalysts or Enzymes

Processes, substrates, end products, temperature stability, and enzyme source can all be classified. Amylases (starch-degrading enzymes) are commonly used in baking to degrade gelatinizing starches. The stability of temperature and mode of action are critical for long-term use and performance. Amylases are required to change the gelatinization of starch when baking at temperatures above 150 °C, resulting in a soft crumb. The most common amylases are from *Bacillus subtilis*, *Aspergillus niger*, malted wheat/barley, *Bacillus subtilis*, *Bacillus megaterium*, and *Bacillus stearothermophilus*.

9.9.4 Hydrocolloids

Hydrocolloids improve water solubility, viscosity, and the ability to form gels. Ice crystallization is slowed by stable emulsions made with hydrocolloids, which improve and stabilize the texture. Gums are high-molecular-weight polysaccharides that are high in fiber and protein (Kanatt, 2020). Carrageenan, alginates, guar, locust bean gum, and xanthan gum are all common gums in baked goods, as are alginates and alginates. Gums, which are used in icings and meringues, prevent syneresis by binding water and provide freeze/thaw stability with a glossy sheen.

Figure 9.13: Image of hydrocolloids.

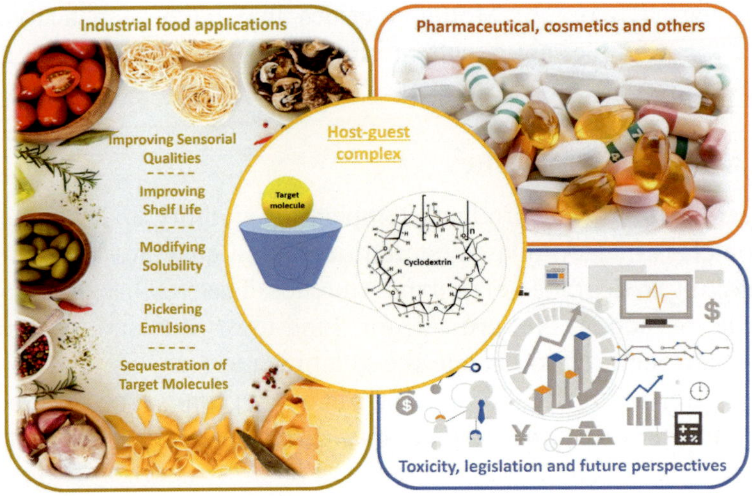

Figure 9.14: Hydrocolloids. (Source: Gonzalez Pereira A, Carpena M, et, al., 2021)

The Movement of Water (Aw)

The abbreviation "Aw" denotes the amount of available moisture in a food system. This is not the same as the total amount of water delivered in the meal. Humectants can be used to control microbiological and chemical activity concerning relative vapor pressure. Water is bound by hydrophilic binding sites in sugars, starches, and salt. Sugars and sugar alcohols such as mannitol, maltose, sucrose, sorbitol, and fructose are the most common humectants (Kuijt, 2009). Because of its versatility, honey is an excellent humectant for frozen dough products. Furthermore, studies have shown that honey's antioxidant properties improve the flavor and texture of food.

Reducing the Number of Microorganisms

Factory health and safety policies aid in the prevention of contamination. In addition to lowering Aw, chemicals can help with microbial management. The low pH of vinegar and concentrated raisin juice inhibits the growth of microorganisms. Betaine and propionate are effective antifungals, antimicrobial, and antibacterial agents in food.

Innovative Food Preservation Techniques

Protein Because of their ability to bind water, soy, whey, and milk components can help meals last longer. Baked proteins contain additional water-binding sites that were not present in the original protein form. When these ingredients are used in bread filler recipes, they produce a moist final product with low starch retrogradation. They have a strong affinity for water and bind it tightly to the protein. Changing the moisture level of a product may result in a flavor change. Oil-based flavorings should be used instead of water-based flavorings because they last longer and bake at a lower temperature (Kendall & Dimond, 2007).

Edible Films and Coatings have made it possible to create high-quality meals with a long shelf life. Single-use packaging should be reduced due to environmental concerns, and recycling options should be expanded. Edible coatings improve the mechanical properties of food and reduce volatile flavor loss as a result. Proteins, lipids, and polysaccharides have all been demonstrated to be effective mass transfer barriers. Refrigeration does not preserve foods in the same way that freezing does. Staling has a negative impact on food flavor and texture. This effect can be reduced by freezing. Freeze your baked goods in an airtight container to ensure that they don't go to waste.

Packages of Excellent Quality

Packaging is important when it comes to baked goods with a shelf life of two to six months or more. Barrier films of good quality are widely available at reasonable prices. Researchers discovered that using modified-atmosphere packaging (MAP), which incorporates gases to manage the environment, can increase the shelf life of some foods by up to a week. A lack of seals, packaging integrity, and residual oxygen in the pack may result in unsafe baked dinners. Bacterial growth in anaerobic conditions is possible even when the area is properly sealed (Galić et al., 2009). Manufacturers should monitor Aw and pH levels to prevent the growth of these microbes. Because of shelf life, customers should be able to make well-informed and safe product choices. Consumers must understand when a product is safe to consume. Opened items must be stored by certain guidelines, such as temperature and time limits.

CHAPTER 10

FOOD PRESERVATION, SAFETY, AND SHELF LIFE EXTENSION

CONTENTS

Food preservation is related to food safety and also the shelf life of the product. How food is preserved will determine whether it can last for long without spoiling. Poor food preservation will shorten the shelf life of the product. This is because the product can spoil before the indicated expiry date. Food preservation is a combination of various practices done to food to ensure that it can last for long periods. It includes food processing practices that are useful in preventing the growth of various microorganisms such as yeast and slowing down the oxidation of fats that cause rancidity (Howey, 2015). Even though food preservation seeks to prevent the growth of microorganisms, some of the methods used in preserving food work by introducing benign bacteria to fungi into food. There are cases where food preservation includes processes that inhibit visual deterioration. This includes the enzymatic browning reaction noted in apples after they are sliced or cut during food preparation. Food preservation is useful in ensuring minimum amounts of food are wasted. It is also very useful in decreasing production costs while increasing the efficiency of food systems, improving food security and the needed nutrition that contributes to environmental sustainability (Hedegaard & Skibsted, 2013). A good example is when food preservation is used in reducing the environmental impact of food production.

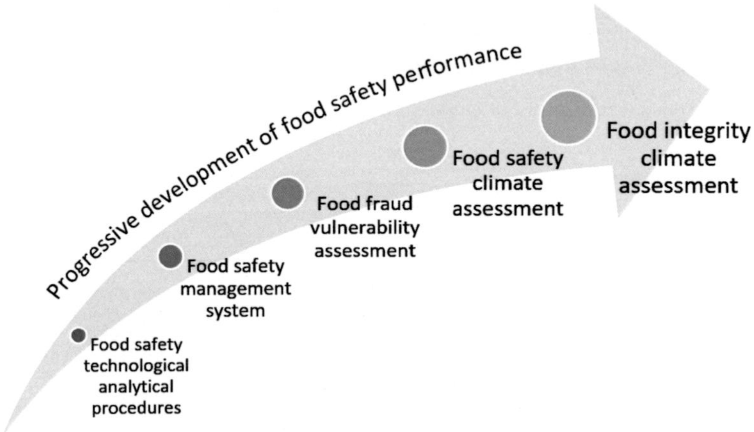

Figure 10.1: Food preservation, safety, and shelf life extension are inter-related. (Source: Alrobaish WS, Jacxsens L, *et, al.,*2021)

Various processes have been tailored to preserve food. In most cases, food processes are made of a combination of other food preservation methods. In the case of fruit preservation conversion into jam then it involves boiling. Boiling is done to reduce the amount of moisture in food. It is also done to kill bacteria. Sugaring is also done to prevent regrowth while sealing with an air-tight jar is done to prevent recontamination. Food preservation method have different effects on the quality of food and food systems. Studies on good preservation methods show that traditional methods of preserving food items that have been shown to have a lower carbon footprint and energy input compared to some modern methods (GÁMbaro et al., 2006). Some methods of food preservation are a threat to human health as they are known to create carcinogens. For instance, the International Agency for Research on Cancer in 2015 classified processed meat that undergoes salting, curing, fermenting and smoking as carcinogenic to human beings. There are two main classifications of food preservation methods namely traditional and modern. Some traditional methods are used currently though they have been upgraded.

10.1 TRADITIONAL METHODS

Some of the methods used in food preservation can date to the dawn of agriculture while others were discovered recently. Some of the traditional methods of food preservation are discussed as follows:

10.1.1 Boiling

This method of food preservation works by killing existing microbes. The most common types of food preserved through this method are milk and water. They are both boiled to kill any harmful microbes in them.

10.1.2 Burial

This method of food preservation is used as it can protect food from a variety of factors such as lack of oxygen, lack of light, pH levels, desiccants, and cool temperatures in soil. This method of food preservation is often used alongside other methods of preservation that include fermentation and salting. Food preservation through burial is done with soil that is very dry and salty such as sand. Frozen soil can also be used for such kind of food preservation. Root vegetables, are known to be very resistant to spoilage and therefore need no other preservation than storage in cool dark conditions (Hosseini & Jafari, 2020). A good example is a burial in the ground such as a storage clamp often confused with a root cellar. Traditionally, cabbage was preserved by burying it during autumn. There are methods used to ensure that the cabbage is crispy while other methods produce sauerkraut. The latter is also used in the production of kimchi.

There are cases where meat is buried under conditions that cause preservation. Meat can be buried in hot ashes or coals. The heat in this case kills pathogens with dry ash desiccate while the earth blocks oxygen and further contamination. When meat is buried when the earth is very cold then the earth acts as a refrigerator. In areas that experience permafrost, the earth acts as a freezer. In some parts of India, it is more practical to store rice through underground burying. This method has proven useful in storing rice for three to six months during the dry season. Other forms of preservation include butter and other similar substances being preserved as bog butter. The placement of eggs in alkaline mud leads to the creation of century eggs. This results in inorganic fermentation through raised pH. The fermentation process is useful in preserving them while breaking down some of the complex, less flavorful proteins, and fats into simpler and more flavorful ones.

10.1.3 Canning

Canning is a method of food preservation that involves the cooking and storage of food in sterilized jars or cans. Containers used in canning are boiled to kill or weaken any form of remaining Bacteria as a form of sterilization. Traditionally, canning was used in the preservation of meat, vegetables, fruits, and milk. Canning was greatly used when the relationship between microorganisms, food spoilage, and illness was established. Naturally, goods have different degrees of natural protection against spoilage. Such food may at times require a final step that occurs in a pressure cooker. Preservation of high-acid fruits such as strawberries may not require the use of preservatives. Strawberries may need a short boiling cycle while marginal vegetables like the likes of carrots may need longer boing times and the addition of other acidic elements (Galovičová et al., 2021). Vegetables and meat are low-acid foods that need pressure canning. Compared to other methods of food preservation z preservation through canning and boiling is at a larger risk of spoilage once the bottle or can has been opened. The canning process needs quality control and failure to do so will allow ingress of water or microorganisms.

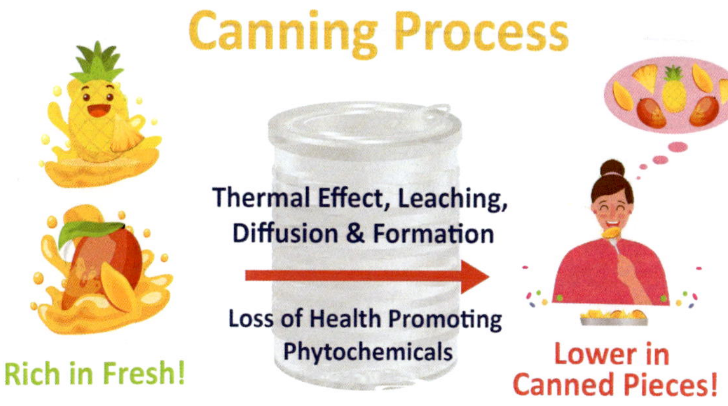

Figure 10.2: Cans used in storage are sterilized before use. (Source: Arampath PC, Dekker M. *et, al.,* 021)

Failures in canning can rapidly be detected as decomposition within the can. Decomposition is known to lead to gas production leading to the bursting or swelling of the can. Poor manufacture of can foods is though under processing. When canning foods, one has to ensure that high levels of hygiene are maintained. Any form of poor hygiene will lead to food contamination by the obligate anaerobe *Clostridium botulinum*. It leads to the production of an acute toxin within the good. When such food is consumed, it leads to severe illnesses or death. The challenge with this organism is that it does not produce gas or obvious taste and can remain undetected by smell or taste (He et al., 2018). Such kinds of toxins can be denatured by cooking. For food such as mushrooms, when they are cooked butt handled poorly and canned can support the growth of *Staphylococcus aureus* known to produce a toxin that is not destroyed through canning or subsequent reheating.

10.1.4 Confit

This method of food preservation is mostly used when preserving meat and involves the use of salt or cooking food at 100 °C or near that kind of temperature. When preservation is done through cooking, some kind of fat is used and can be tallow or lard. Storage is done with meat being immersed in fat. This method of preservation was most popular in Europe before the introduction of refrigerators. This method of food preservation is currently used in France. This kind of food preservation and preparation method will keep food for longer durations when stored in a cold cellar or when it is buried in cold ground.

10.1.5 Cooling

Cooling works by slowing down the growth and reproduction of microorganisms and any enzymatic action that can cause food to rot. This method of food preservation has recently been advanced through the introduction of commercial and domestic refrigerators. This has been useful among large populations of people more so in the western world as it allows foods such as dairy products, salads, and fresh fruits to be stored for longer periods more so during warm weather. In the era when mechanical refrigeration was not introduced, food preservation through cooking occurred in the forms of root cellars and iceboxes (Goldman et al., 2002). Traditionally, people would do their own ice cutting while those who lived in the city and twins were dependent on ice trade. Root cellaring is currently used among people for various reasons including heirloom crops, local food, family farming, traditional home cooking techniques, frugality, organic farming and self-sufficiency.

10.1.6 Freezing

Among the most common method of food, preservation is freezing. Freezing is used in commercial and domestic preservation. It can be used to preserve a wide range of food including the preparation of foods that would not have required freezing in their unprepared state. A good example is potato waffles that can be stored in the freezer while potatoes are to be stored in a cool dark place to ensure many months of storage (Zhukov et al., 2018). Cold storage is known to provide large-volume, long-term storage for strategic food stocks that have been held by the state in case of a national emergency.

Figure 10.3: Freezing involves food being stored in very low temperatures.

10.1.7 Heating

Heating is useful in food preservation as food is subjected to very high temperatures that are sufficient to kill microorganisms inside the food. It is the same method used with perpetual stews.

10.1.8 Jellying

In this method of food preservation, food is preserved by cooking it in a material that solidifies forming a gem. Some of the materials used in solidification include arrowroot flour, maize flour agar, and gelatin. Protein gel can bc formed when animal flesh is preserved through jellying. In Xiamen China, sipunculid worms, eels and elvers are considered a delicacy. This is the case for jellied eels that are eaten with mashed potatoes. In the 1950s meat off-cuts were preserved in aspic which is a gel made from gelatin and clarified meat broth. There are other forms of preservation including the setting of cooked food in containers and covering it with a layer of fat. This method of food preparation and preservation can be used for potted chicken liver. The same applies to potted shrimps that can be served on toast. Invalids were served with calf's foot jelly. Jellying of food is very essential in the production of traditional pâtés. There are different kinds of jugged meats that have been jellied (Weinbreck et al., 2010).

Jellying has also been used in the preservation of fruits. Preparations involve the use of cooked fruits, sugar and vegetables. After preparation, it is stored in glass jam jars and Mason jars there are different kinds of food preserved made worldwide including fruit preserves such as those made from apricot, strawberry and savory preserves. They also include those made from tomatoes or squash. There are various factors that influence the type of preserves which include the type of ingredients bused and the method of preparation. Some examples of preserves are marmalades, jellies and jams all of which are different styles of fruit preserves that vary on the basis of the type of fruit used.

10.1.9 Jugging

This method of preservation is mostly used in preserving meat. This process involves the stewing of meat more so game or fish in a casserole or covered earthenware jug. Preparation involves the animal being jugged being cut into pieces and placed into a tightly sealed jug with gravy or brine and stewed.

The animal's blood or red wine is at times added to the cooking liquid. This method of food preservation was very popular in the 20 century.

10.1.10 Lye

This method of food preservation works by preventing the growth of bacteria. It mostly involves the use of Sodium hydroxide that makes food too alkaline limiting the survival of bacteria. Lye can saponify fats in food leading to a change in its texture and flavor. Lye is used in Lutefisk during its preparation. The same applies to some olive recipes. Lye is also used in modern recipes of century eggs (Villalobos-Delgado et al., 2019).

10.1.11 Pickling

This method of food preservation involves food being placed in an edible antimicrobial liquid. Pickling can be subdivided into two categories namely fermentation pickling and chemical pickling. Under chemical pickling, food is placed in an edible liquid that kills bacteria or inhibits its growth. Some of the most common pickling agents include brine, vegetable oil, alcohol and vinegar. Boiling or heating are some of the chemical pickling processes used in pickling so that preserved food is saturated with the pickling agent. Eggs, herring, corned beef, peppers and cucumbers are some of the common chemically pickled foods. This is very different from fermentation pickling in that the bacteria in the liquid produce organic acids as their preservation agents. The preservation agents are produced through lactic acid through the presence of Lactobacillus. Kimchi, nukazuke, sauerkraut and surströmming are examples of fermented pickles.

10.1.12 Sugaring

Sugar is the earliest form of culture that has been used as a preservative. A good example was when food was placed in honey to preserve it. The method of preservation is similar to the preservation of pickled foods. Sugaring was greatly used in areas that experience northern climates that do not experience sufficient sun to dry food. Preservation was done through the heating of fruit with sugar. Naturally, sugar tends to draw water from the microbes. The result of the process is that the microbial cells are left dehydrated. Dehydration in such cells kills them. The process leaves food safe from any form of microbial spoilage (Wrona & Nerín, 2020).

Sugar is mostly used for fruit preservation either in a crystallized form where the preserved material is cooked in sugar till it reaches the point of

Crystallization and the final product is stored dry or in an antimicrobial syrup with fruits such as plums, apricots, peaches, pears and apples. Sugaring is used for the skins of ginger, angelica and citrus fruits. It can also be used in the production of jelly and jam.

10.2 MODERN INDUSTRIAL METHODS

These methods of food preservation are those developed in research laboratories for commercial applications. They are discussed as follows:

10.2.1 Vacuum Packing

This method of food preservation involves food being placed in a vacuum environment. The vacuum environment may be an air-tight bag or bottle. This kind of environment is useful in stripping bacteria of oxygen needed to ensure their survival. This method is mostly used in storing nuts to minimize loss of flavor through oxidation. The major challenge of vacuum packaging at the consumer level is that vacuum sealing of products can lead to the deformation of contents while other foods are robbed of their flavor.

Figure 10.4: Vacuum packaging can be used for meat and other products.

10.2.2 Pasteurization

This method of food preservation is used for liquid foods. Initially, it was applied to combat the souring of local wines. This method of preservation for milk will involve it being heated at high temperatures for a given duration of time to kill bacteria present in it and finally it is cooked to prevent the remaining bacteria from growing. The treated milk is then stored in sterilized pouches or bottles and placed in cold places.

10.3 FOOD SAFETY

Food safety defines the routines in preparation, handling, and storage of food done to prevent foodborne illnesses and injury. There are numerous health hazards that good can encounter from farm to factory to fork. This is because there are various points of contamination throughout the supply chain. Safe good handling procedures and practices are implemented at every stage of the food production life cycle to curb the risks and prevent harm to consumers (Winterhalder et al., 2015). Food safety is a scientific discipline that draws from a variety of academic fields such as chemistry, engineering, and microbiology. The diverse schools of thought converge intending to ensure that food processing safety is carried out wherever food products are obtained, manufactured, prepared, stored, and sold. For this reason, food safety can be viewed as a systematic approach to hygiene and accountability that concerns every aspect of the global food industry.

Across the world, there are food safety regulations to help govern food production across the world. Some of the regulations are similar between countries as some food products are exported and imported by countries. This is because most food products are the most-traded commodities. In the world today, markets are becoming more globalized and with the increasing growth of the world's population, the global food supply chain is expected to increase in scale and complexity. This is attributed to the fact that the megatrends are known to influence the mass production of food and also its distribution placing great importance on food safety compliance. Every state has its regulatory bodies that preside over the definition and also enforce domestic food safety standards. For a food company to sell their products bin another country, their products are usually subjected to food safety legislation and enforcement measures of that country (Wibowo et al., 2018). A good example is the European Union which has been involved in food safety legislation that has detailed its regulations. In the case of the United States, the legal requirements for food safety are outlined in the Food Safety Modernization Act. Across the world, it noted that laws of food safety are formed on the basis of two concepts namely GMP and HACCP.

Figure 10.5: High levels of hygiene are a good manufacturing practice. (Source: Zhong C, Reibman AR., *et, al*.,2020)

GMP stands for Good Manufacturing Practices. These practices are usually recognized internationally as quality assurance guidelines for the production of beverages, goods and dietary supplements among others. The guidelines provide a layout of protocols that must be implemented by manufacturers to assure that their products are consistently high quality while ensuring that they are safe for human consumption. Some of the practices include mandatory product inspection at critical control points. HACCP on the other hand stands for Hazard Analysis and Critical Control Points. It is considered a systematic, risk-based approach to preventing the chemical, biological and physical contamination of food in production, packaging and distribution (Uebele et al., 2013). This concept was designed to counter health hazards through the identification of potential food safety problems before they occur rather than inspecting food products for hazards after the fact. The concept also entails the control of contaminants at a number of key junctures in food production processes. There is also a high adherence to hygiene practices.

Some companies adopt the use of privately-owned international organizations that provide comprehensive guidelines for auditing food manufacturers on the basis of hygiene and good safety. Setting international standards is useful in facilitating global food trade by assisting food industry players from different countries ensuring food safety and quality are met at the borders. Countries are required to comply with food safety laws in countries they are active. The global market leaders in the food industry tend to pursue certification with a number of private food regulators. Furthermore, there is a demand that the downstream and upstream suppliers they work with provide proof of similar certification.

For the international regulatory bodies, there are various people responsible for food safety standards. These bodies are useful in providing guidance, certification and auditing services for food manufacturers globally. It is important to note that they are not responsible for the active enforcement of food safety laws. Laws on food safety are defined by the nation. They also establish and enforce the laws for food safety regulations. There may be a variation in food regulations from country to country or domestically (Turner et al., 2020). For a company to bring food products to a foreign market, they need to be compliant with food safety and consumer protection laws of the nation and its regional government authorities. When ensuring food safety, there are seven principles of a safe food supply chain.

HACCP helps by laying out the steps essential in proactively ensuring food quality safety in individual food production environments. This is very vital in a healthy supply chain which demands action on a collective level. There are seven overarching food safety principles that have been identified by the European Union as necessary for the entire food supply chain to operate for the good of the public. The principles include corporate responsibility where by every company in the food supply chain needed to perform their due diligence. This is done with the aim of ensuring the quality and safety of food products within the bounds of their responsibility. Such kind of responsibility includes the implementation of in-house controls according to HACCP. It is very essential for corporations to assume liability for any damages caused by their products.

The second principle is traceability where all food business operators in the European Union have the responsibility of documenting the source of raw materials and where they are sent. Regulatory bodies make use of this documentation when they want to identify the source of contamination more so when there is a need for a recall. Official food control is the third principle. It involves government authorities within the federal states being tasked with the responsibility of enforcing food law requirements through targeted sample collection, risk-oriented reviews, and regular inspections.

The fourth principle of the precautionary principle involves competent authorities being permitted to take precautionary measures when they believe the effect will help reduce food safety risks. The precautionary measures are reviewed on an ongoing basis when scientific data becomes available. The fifth principle is independent scientific risk assessment. This principle is used when a governmental institution operating independently of economy, social and political influence being responsible for any scientific investigation and assessing the risks linked to food products. The next principle is the separation of risk assessment and risk management. This was brought about by conflicts of interest. In such as case, a clear distinction is made between those responsible for risk management and those responsible for scientific risk assessment (Amani & Gadde, 2015). The final principle is transparent risk communication. This principle requires the public being promptly informed if potential and imminent food safety hazards. For information on food safety to be circulated, scientists in conjunction with food business operators and policymakers should communicate transparently.

10.4 EXTENSION OF FOOD SHELF LIFE

Extended food shelf life is useful in dealing with problems of food security. The method used in extending food shelf life will determine whether food is safe for consumption. Extension of food shelf life use useful in that fresh produce can spend up to half its life in transit. For it to reach its final destination, efforts have to be put in place. This gives products the best chance of survival in long-haul shipments. There are numerous benefits of extended food shelf life. They include better brand reputation, increased margins, less food loss, and wastage, fewer shipments rejected and better quality produce. There are different approaches taken by food production companies to extend food shelf life. They include several newly developed innovations.

To improve food shelf life, one needs to know what causes reduced food shelf life. Some of the factors affecting food shelf life include physical harm, microbial growth, moisture or humidity, temperature changes and extremes, and finally ethylene exposure. There are different methods used by companies to extend food shelf life. The methods vary depending on the type of product (Timakova et al., 2020). Some of the methods include packaging improvements and modified atmosphere packaging, monitoring humidity, finding weaknesses in the cold chain, improving treatment and handling procedures, use of natural antimicrobials, utilization of new varieties of crops and ingredients, absorbent food trays and pads and use of smart stickers. Appropriate methods of food shelf life extension should be used to minimize any threat to human health.

REFERENCES

1. Alkanan ZT, Altemimi AB, Al-Hilphy ARS, Watson DG, Pratap-Singh A. Ohmic Heating in the Food Industry: Developments in Concepts and Applications during 2013–2020. *Applied Sciences*. 2021; 11(6):2507.
2. Alrobaish WS, Jacxsens L, Luning PA, Vlerick P. Food Integrity Climate in Food Businesses: Conceptualization, Development, and Validation of a Self-Assessment Tool. *Foods*. 2021; 10(6):1302.
3. Amani, P., & Gadde, L. E. (2015). Shelf life extension and food waste reduction. *Proceedings in Food System Dynamics*, 7-14.
4. Angourakis, A., Santos, J. I., Galán, J. M., & Balbo, A. L. (2015). Food for all: An agent-based model to explore the emergence and implications of cooperation for food storage. *Environmental Archaeology*, *20*(4), 349-363.
5. Balakrishnan N, Yusop SM, Rahman IA, Dauqan E, Abdullah A. Efficacy of Gamma Irradiation in Improving the Microbial and Physical Quality Properties of Dried Chillies (*Capsicum annuum* L.): A Review. *Foods*. 2022; 11(1):91.
6. Calligaris, S., Manzocco, L., Anese, M., & Nicoli, M. C. (2016). Shelf-life assessment of food undergoing oxidation–a review. *Critical reviews in food science and nutrition*, *56*(11), 1903-1912.
7. Castrica, M., Chiesa, L. M., Nobile, M., De Battisti, F., Siletti, E., Pessina, D.,& Balzaretti, C. M. (2021). Rapid safety and quality control during fish shelf-life by using a portable device. *Journal of the Science of Food and Agriculture*, *101*(1), 315-326.

8. Chang, K. H., Hurley, B. L., Pascall, M. A., & Frankel, G. S. (2022). Corrosion in Tinplate Cans Used for Food Storage, Part 3: Effects of Cysteine, NaCl, and Tomatoes on Corrosion of Tin, Iron, and Tinplate. *Corrosion*, *78*(2), 127-141.

9. Chiang JH, Henry CJ. Plant Proteins for Future Foods: A Roadmap. *Foods*. 2021; 10(8):1967.

10. Coffigniez F, Matar C, Gaucel S, Gontard N, Guilbert S and Guillard V (2021) The Use of Modeling Tools to Better Evaluate the Packaging Benefice on Our Environment. *Front. Sustain. Food Syst.* 5:634038.

11. Dawson, P., Al-Jeddawi, W., & Remington, N. (2018). Effect of freezing on the shelf life of salmon. *International Journal of Food Science, 2018*.

12. De Corato, U. (2020). Improving the shelf-life and quality of fresh and minimally-processed fruits and vegetables for a modern food industry: A comprehensive critical review from the traditional technologies into the most promising advancements. *Critical Reviews in Food Science and Nutrition*, *60*(6), 940-975.

13. de Oliveira Filho JG, Miranda M, Ferreira MD, Plotto A. Nanoemulsions as Edible Coatings: A Potential Strategy for Fresh Fruits and Vegetables Preservation. *Foods*. 2021; 10(10):2438.

14. Derkach, S. R., Voron'ko, N. G., Kuchina, Y. A., & Kolotova, D. S. (2020). Modified fish gelatin as an alternative to mammalian gelatin in modern food technologies. *Polymers*, *12*(12), 3051.

15. Drago E, Campardelli R, Pettinato M, Perego P. Innovations in Smart Packaging Concepts for Food: An Extensive Review. *Foods*. 2020; 9(11):1628.

16. El-Anany, A. M., Hassan, G. F. A., & Ali, F. R. (2009). Effects of edible coatings on the shelf-life and quality of Anna apple (Malus domestica Borkh) during cold storage. *Journal of food technology*, *7*(1), 5-11.

17. Fan, W., Sun, J., Chen, Y., Qiu, J., Zhang, Y., & Chi, Y. (2009). Effects of chitosan coating on quality and shelf life of silver carp during frozen storage. *Food chemistry*, *115*(1), 66-70.

18. Fiorentini M, Kinchla AJ, Nolden AA. Role of Sensory Evaluation in Consumer Acceptance of Plant-Based Meat Analogs and Meat Extenders: A Scoping Review. *Foods*. 2020; 9(9):1334.

19. Galić, K., Ćurić, D., & Gabrić, D. (2009). Shelf life of packaged bakery goods—A review. *Critical reviews in food science and nutrition*, *49*(5), 405-426.

20. GÁMbaro, A., Ares, G., & Gimenez, A. N. A. (2006). Shelf-life estimation of apple-baby food. *Journal of sensory studies*, *21*(1), 101-111.

21. Galovičová, L., Borotová, P., Valková, V., Vukovic, N. L., Vukic, M., Terentjeva, M., ... & Kačániová, M. (2021). Thymus serpyllum Essential Oil and Its Biological Activity as a Modern Food Preserver. *Plants*, *10*(7), 1416.

22. Giannakourou, M., & Taoukis, P. (2020). Holistic Approach to the Uncertainty in Shelf Life Prediction of Frozen Foods at Dynamic Cold Chain Conditions. *Foods (Basel, Switzerland)*, *9*(6), 714.

23. Goldman, A., Ramaswami, S., & Krider, R. E. (2002). Barriers to the advancement of modern food retail formats: theory and measurement. *Journal of Retailing*, *78*(4), 281-295.

24. Gonzalez Pereira A, Carpena M, García Oliveira P, Mejuto JC, Prieto MA, Simal Gandara J. Main Applications of Cyclodextrins in the Food Industry as the Compounds of Choice to Form Host–Guest Complexes. *International Journal of Molecular Sciences*. 2021; 22(3):1339.

25. Haji M, Kerbache L, Muhammad M, Al-Ansari T. Roles of Technology in Improving Perishable Food Supply Chains. *Logistics*. 2020; 4(4):33.

26. He, T., Wang, H., Chen, Z., Liu, S., Li, J., & Li, S. (2018). Natural quercetin AIEgen composite film with antibacterial and antioxidant properties for in situ sensing of Al3+ residues in food, detecting food spoilage, and extending food storage times. *ACS Applied Bio Materials*, *1*(3), 636-642.

27. Hedegaard, R. V., & Skibsted, L. H. (2013). Shelf-life of food powders. In *Handbook of food powders* (pp. 409-434). Woodhead Publishing.

28. Hosseini, H., & Jafari, S. M. (2020). Introducing nano/microencapsulated bioactive ingredients for extending the shelf-life of food products. *Advances in Colloid and Interface Science*, *282*, 102210.

29. Howey, M. C. (2015). Geospatial landscape permeability modeling for archaeology: a case study of food storage in northern Michigan. *Journal of Archaeological Science*, *64*, 88-99.

30. Hsu H-C, Liao P-W, Lee H-T, Liu W-C, Ho M-L. Silver Nanoplates for Colorimetric Determination of Xanthine in Human Plasma and in Fish Meat via Etching/Aggregation/Fusion Steps. *Sensors*. 2020; 20(20):5739.

31. Ibrahim SA, Ayivi RD, Zimmerman T, Siddiqui SA, Altemimi AB, Fidan H, Esatbeyoglu T, Bakhshayesh RV. Lactic Acid Bacteria as Antimicrobial Agents: Food Safety and Microbial Food Spoilage Prevention. *Foods*. 2021; 10(12):3131.
32. Kendall, P., & Dimond, N. (2007). Food storage for safety and quality. *Food and nutrition series. Health; no. 9.310*.
33. Kuijt, I. (2009). What do we really know about food storage, surplus, and feasting in preagricultural communities?. *Current Anthropology*, *50*(5), 641-644.
34. Kanatt, S. R. (2020). Development of active/intelligent food packaging film containing Amaranthus leaf extract for shelf life extension of chicken/fish during chilled storage. *Food Packaging and Shelf Life*, *24*, 100506.
35. Lee JC, Daraba A, Voidarou C, Rozos G, Enshasy HAE, Varzakas T. Implementation of Food Safety Management Systems along with Other Management Tools (HAZOP, FMEA, Ishikawa, Pareto). The Case Study of *Listeria monocytogenes* and Correlation with Microbiological Criteria. *Foods*. 2021; 10(9):2169.
36. Łepecka A, Szymański P, Rutkowska S, Iwanowska K, Kołożyn-Krajewska D. The Influence of Environmental Conditions on the Antagonistic Activity of Lactic Acid Bacteria Isolated from Fermented Meat Products. *Foods*. 2021; 10(10):2267.
37. Liu, L., Wang, J., Rosenberg, D., Zhao, H., Lengyel, G., & Nadel, D. (2018). Fermented beverage and food storage in 13,000 y-old stone mortars at Raqefet Cave, Israel: Investigating Natufian ritual feasting. *Journal of Archaeological Science: Reports*, *21*, 783-793.
38. Li E, Saleem F, Edge TA, Schellhorn HE. Biological Indicators for Fecal Pollution Detection and Source Tracking: A Review. *Processes*. 2021; 9(11):2058
39. Lindström, M., Kiviniemi, K., & Korkeala, H. (2006). Hazard and control of group II (non-proteolytic) *Clostridium botulinum* in modern food processing. *International journal of food microbiology*, *108*(1), 92-104.
40. Martins, R. C., Lopes, V. V., Vicente, A. A., & Teixeira, J. A. (2008). Computational shelf-life dating: complex systems approaches to food quality and safety. *Food and Bioprocess Technology*, *1*(3), 207-222.

41. Malik H, Afthanorhan A, Amirah NA, Fatema N. Machine Learning Approach for Targeting and Recommending a Product for Project Management. Mathematics. 2021; 9(16):1958.

42. Mahmud N, Islam J, Tahergorabi R. Marine Biopolymers: Applications in Food Packaging. *Processes*. 2021; 9(12):2245.

43. Manzocco, L., Calligaris, S., & Nicoli, M. C. (2010). Methods for food shelf life determination and prediction. In *Oxidation in foods and beverages and antioxidant applications* (pp. 196-222). Woodhead Publishing.

44. Maertens, M., Minten, B., & Swinnen, J. (2012). Modern food supply chains and development: Evidence from horticulture export sectors in Sub-Saharan Africa. *Development Policy Review*, *30*(4), 473-497.

45. Melini V, Melini F. Strategies to Extend Bread and GF Bread Shelf-Life: From Sourdough to Antimicrobial Active Packaging and Nanotechnology. Fermentation. 2018; 4(1):9.

46. Mustafa F, Andreescu S. Chemical and Biological Sensors for Food-Quality Monitoring and Smart Packaging. *Foods*. 2018; 7(10):168

47. Nagi, H. P. S., Kaur, J., Dar, B. N., & Sharma, S. (2012). Effect of storage period and packaging on the shelf life of cereal bran incorporated biscuits. *American Journal of Food Technology*, *7*(5), 301-310.

48. Nardella S, Conte A, Del Nobile MA. State-of-Art on the Recycling of By-Products from Fruits and Vegetables of Mediterranean Countries to Prolong Food Shelf Life. *Foods*. 2022; 11(5):665.

49. Njage PMK, Leekitcharoenphon P, Hansen LT, Hendriksen RS, Faes C, Aerts M, Hald T. Quantitative Microbial Risk Assessment Based on Whole Genome Sequencing Data: Case of *Listeria*

50. monocytogenes. *Microorganisms*. 2020; 8(11):1772.

51. Nirupama, P., Gol, N. B., & Rao, T. R. (2010). Effect of post harvest treatments on physicochemical characteristics and shelf life of tomato (Lycopersicon esculentum Mill.) fruits during storage. *American-Eurasian Journal of Agricultural & Environmental Sciences*, *9*(5), 470-479.

52. Nicoli, M. C. (2012). The shelf life assessment process. *Shelf Life Assessment of Food, Nicoli; Nicoli, MC, Ed.; CRC PRESS, Taylor & Francis Group: Boca Raton, FL, USA*, 17-36.

53. O'Brien, M. J., & Bentley, R. A. (2015). The role of food storage in human niche construction: An example from Neolithic Europe. *Environmental archaeology*, *20*(4), 364-378.
54. Omotilewa, O. J., Ricker-Gilbert, J., Ainembabazi, J. H., & Shively, G. E. (2018). Does improved storage technology promote modern input use and food security? Evidence from a randomized trial in Uganda. *Journal of Development Economics*, *135*, 176-198.
55. Peleg, M., Normand, M. D., & Corradini, M. G. (2017). A new look at kinetics in relation to food storage. *Annual review of food science and technology*, *8*, 135-153.
56. Paiva T, Jacinto TA, Sarraguça MC, Coutinho P. Beef Consumers Behaviour and Preferences—The Case of Portugal. *Sustainability*. 2022; 14(4):2358.
57. Pratiwi, A. S., Dwivany, F. M., Larasati, D., Islamia, H. C., & Martien, R. (2015, September). Effect of chitosan coating and bamboo FSC (fruit storage chamber) to expand banana shelf life. In *AIP Conference Proceedings* (Vol. 1677, No. 1, p. 100005). AIP Publishing LLC.
58. Rasmussen EDJ, Errico M, Tronci S. Adaptive Feedback Control for a Pasteurization Process. *Processes*. 2020; 8(8):930.
59. Rathod NB, Nirmal NP, Pagarkar A, Özogul F, Rocha JM. Antimicrobial Impacts of Microbial Metabolites on the Preservation of Fish and Fishery Products: A Review with Current Knowledge. *Microorganisms*. 2022; 10(4):773.
60. Roccato, A., Uyttendaele, M., & Membré, J. M. (2017). Analysis of domestic refrigerator temperatures and home storage time distributions for shelf-life studies and food safety risk assessment. *Food Research International*, *96*, 171-181.
61. Sarkar, T., Salauddin, M., Pati, S., Chakraborty, R., Shariati, M. A., Rebezov, M.,& Lorenzo, J. M. (2022). The fuzzy cognitive map–based shelf-life modeling for food storage. *Food Analytical Methods*, *15*(3), 579-597.
62. Schmidt M, Zannini E, Arendt EK. Recent Advances in Physical Post-Harvest Treatments for Shelf-Life Extension of Cereal Crops. *Foods*. 2018; 7(4):45.
63. Shen C-Y, Lu C-H, Wu C-H, Li K-J, Kuo Y-M, Hsieh S-C, Yu C-L. The Development of Maillard Reaction, and Advanced Glycation End Product (AGE)-Receptor for AGE (RAGE) Signaling Inhibitors

as Novel Therapeutic Strategies for Patients with AGE-Related Diseases. Molecules. 2020; 25(23):5591

64. Sim SYJ, SRV A, Chiang JH, Henry CJ. Plant Proteins for Future Foods: A Roadmap. *Foods*. 2021; 10(8):1967.
65. Sonwani E, Bansal U, Alroobaea R, Baqasah AM and Hedabou M (2022) An Artificial Intelligence Approach Toward Food Spoilage Detection and Analysis. *Front. Public Health* 9:816226.
66. Sumner, J., & Jenson, I. (2011). The effect of storage temperature on shelf life of vacuum-packed lamb shoulders. *Food Australia*, *63*(6), 249-251.
67. Sewald, M., & DeVries, J. (2003). Food product shelf life. *Medallion Laboratories Analytical Progress*, 1-10.
68. Sutton, A. O., Strickland, D., & Norris, D. R. (2016). Food storage in a changing world: implications of climate change for food-caching species. *Climate Change Responses*, *3*(1), 1-25.
69. Tagarakis AC, Benos L, Kateris D, Tsotsolas N, Bochtis D. Bridging the Gaps in Traceability Systems for Fresh Produce Supply Chains: Overview and Development of an Integrated IoT-Based System. *Applied Sciences*. 2021; 11(16):7596
70. Timakova, R. T., Tikhonov, S. L., & Tikhonova, N. V. (2020). Ionizing radiation treatment as an innovative process approach in food storage technology for modern agriculture. In *IOP Conference Series: Earth and Environmental Science* (Vol. 421, No. 2, p. 022015). IOP Publishing.
71. Turner, E. R., Luo, Y., & Buchanan, R. L. (2020). Microgreen nutrition, food safety, and shelf life: A review. *Journal of food science*, *85*(4), 870-882.
72. Torres-Sánchez R, Martínez-Zafra MT, Castillejo N, Guillamón-Frutos A, Artés-Hernández F. Real-Time Monitoring System for Shelf Life Estimation of Fruit and Vegetables. *Sensors*. 2020; 20(7):1860.
73. Uebele, M., Grünebaum, T., & Kopsidis, M. (2013). King's law and food storage in Saxony, c. 1790-1830. *Center for Quantitative Economics Working Papers*, *26*.
74. Verni M, Rizzello CG and Coda R (2019) Fermentation Biotechnology Applied to Cereal Industry By-Products: Nutritional and Functional Insights. *Front. Nutr*. 6:42.

75. Wibowo, S., Buvé, C., Hendrickx, M., Van Loey, A., & Grauwet, T. (2018). Integrated science-based approach to study quality changes of shelf-stable food products during storage: A proof of concept on orange and mango juices. *Trends in Food Science & Technology*, *73*, 76-86.

76. Winterhalder, B., Puleston, C., & Ross, C. (2015). Production risk, inter-annual food storage by households and population-level consequences in seasonal prehistoric agrarian societies. *Environmental Archaeology*, *20*(4), 337-348.

77. Wohner B, Pauer E, Heinrich V, Tacker M. Packaging-Related Food Losses and Waste: An Overview of Drivers and Issues. *Sustainability*. 2019; 11(1):264..

78. Wrona, M., & Nerín, C. (2020). Analytical approaches for analysis of safety of modern food packaging: A review. *Molecules*, *25*(3), 752.

79. Villalobos-Delgado, L. H., Nevárez-Moorillon, G. V., Caro, I., Quinto, E. J., & Mateo, J. (2019). Natural antimicrobial agents to improve foods shelf life. In *Food quality and shelf life* (pp. 125-157). Academic Press.

80. Verheyen D, Van Impe JFM. The Inclusion of the Food Microstructural Influence in Predictive Microbiology: State-of-the-Art. *Foods*. 2021; 10(9):2119.

81. Verni M, Rizzello CG and Coda R (2019) Fermentation Biotechnology Applied to Cereal Industry By-Products: Nutritional and Functional Insights. *Front. Nutr*. 6:42.

82. Weinbreck, F., Bodnár, I., & Marco, M. L. (2010). Can encapsulation lengthen the shelf-life of probiotic bacteria in dry products? *International journal of food microbiology*, *136*(3), 364-367.

83. Xiang J, Liu F, Wang B, Chen L, Liu W, Tan S. A Literature Review on Maillard Reaction Based on Milk Proteins and Carbohydrates in Food and Pharmaceutical Products: Advantages, Disadvantages, and Avoidance Strategies. Foods. 2021; 10(9):1998.

84. Zhukov, A. D., Ter-Zakaryan, K. A., Tuchaev, D. U., & Petrovsky, E. S. (2018). Energy-efficient insulation of food storage rooms and vegetable stores. *International Agricultural Journal*, (1), 65-67.

85. Zhong C, Reibman AR, Mina HA, Deering AJ. Multi-View Hand-Hygiene Recognition for Food Safety. *Journal of Imaging*. 2020; 6(11):120.

INDEX

S

T

V

W

Y

FSC
www.fsc.org
MIX
Papier | Fördert gute Waldnutzung
FSC® C083411

Kolibri 360 GmbH
Ferdinand-Jühlke-Straße 7
99095 Erfurt, Deutschland
produktsicherheit@kolibri360.de